METALLOCENE TECHNOLOGY

in Commercial Applications

Dr. George M. Benedikt, Editor

Society of Plastics Engineers

Plastics Design Library

Table of Contents

Preface   vii
    George M. Benedikt

Catalytic Systems

Single-Site Supported Catalysts for Ethylene   1
    A. Muñoz-Escalona, L. Mendez, B. Peña, P. Lafuente, J Sancho,
    W. Michiels, G. Hidalgo and Mᵈ. Fca. Martinez-Nuñez

Synthesis of High-Molecular-Weight Elastomeric Polypropylene with
Half-Titanocene/MAO Catalysts   9
    Oing Wu, Benheng Xie, Meiran Xie, Rui Chen

Precessing Methods

  Blending

Is Metallocene Polyethylene Blend with HDPE More Compatible Than with PP? 17
    D. Rana, K. Cho, B. H. Lee and S. Choe

Spherulite Boundary Strengthening: A New Concept in Polymer Blends   23
    A. Lustiger, CN. Marzinsky and R. R. Mueller

Modification of Syndiotactic Polystyrene   29
    Jörg Kressler, Ralf Thomann

Blends of Ethylene/Styrene Interpolymers and Other Polymers: Benefits in
Applications   37
    C.F. Diehl, M. J. Guest, B. I. Chaudhary, Y. W. Cheung,
    W. R. Van Volkenburgh, B.W. Walther

  Extrusion

Easy Processing Metallocene Polyethylene   47
    Ching-Tai Lue

Effect of Metallocene Polyethylene on Heat Sealing Properties of Low
Density Polyethylene Blends   53
    Juan D. Sierra, María del Pilar Noriega and Tim A. Osswald

Extrusion Operation Window for Filled Metallocene Polyethylenes   61
    María del Pilar Noriega and Tim A. Osswald, Omar A. Estrada

Processing Trends of Metallocene Linear Low Density Polyethylenes and
Their Influence on Single Screw Design   69

K. R. Slusarz, C. A. Ronaghan, and J. P. Christiano
Influence of mPE Grades on the Dynamic Properties of PP/mPE-Blends   77
    Frank Raue and Gottfried W. Ehrenstein
"Single-Site" Catalyzed Polyolefin for Fresh-Cut Produce Packaging.
A Comparison Between Monoextruded Blends and Coextruded Film   85
    V. Patel, S. Mehta, S. A. Orroth and S. P. McCarthy
Rheology and Processing Experience of New Metallocene Polyolefins   93
    Atul Khare, Chuan Qin, Michael T. K. Ling, and L. Woo

Film Forming
Properties and Film Applications of Metallocene-Based Isotactic Polypropylenes 101
    Aiko Hanyu and Rusty Wheat
Phase Structure Characterization and Processing-Structure-Property Relationships
in Linear Low-Density Polyethylene Blown Films   111
    Jianjun Lu, Baiyi Zhao and Hung-Jue Sue
The Relative Influences of Process and Resin Time-Scales on the MD Tear
Strength of Polyethylene Blown Films   121
    Rajendra K. Krishnaswamy and Ashish M. Sukhadia
Metallocene Catalyzed Polyethylene in Blown Film Applications. A Comparison
Between Monoextruded Blended Films and Coextruded Films   129
    C. M. Beagan, G. M. Mc Nally & W. R. Murphy
Trade-offs in Blown Film Processing-Structure-Property Behavior of LLDPE
Type Resins from Chromium, Metallocene and Ziegler-Natta Catalysts   137
    Ashish M. Sukhadia

Injection Molding
Transparent Metallocene-Polypropylene for Injection Molding   147
    J. Rösch, J. R. Grasmeder
Equipment Design Considerations for Processing with Metallocene Resins   153
    Andrew W. Christie

Other Processing Methods
Product and Process Developments in the Nitrogen Autoclave Process for
Polyolefin Foam Manufacture   157
    D. E. Eaves and N. Witten
Properties of Syndiotactic Polypropylene Fibers Produced from Melt Spinning 167
    Mohan Gownder

Melt Spinning of Syndiotactic Polypropylene: Structure Development and
Implications for Oriented Crystallization 177
    Ravi K. Sura, Prashant Desai and A.S. Abhiraman

Properties
New Polyolefins Characterization by Instrumental Analysis 183
    Carmina Gartner and Juan D. Sierra, Roger Avakian
Dynamic Melt Rheometry Used to Study Degradation of Metallocene
Polyethylene 193
    Scott H. Wasserman, George N. Foster, Douglas J. Yacka
Melt Rheology and Processability of Conventional and Metallocene
Polyethylenes 201
    Choon K. Chai
Rheology of Metallocene-Catalyzed Polyethylenes. The Effects of Branching 209
    Sujan E. Bin Wadud and Donald G. Baird
Extensional and Shear Rheology of Metallocene-Catalyzed Polyethylenes 217
    Sujan E. Bin Wadud and Donald G. Baird
A Criterion for the Onset of Gross Polyolefins: Molecular Structure Effect 223
    Seungoh Kim and John. M. Dealy
Relationship Between Structure and Rheology of Constrained Geometry
Catalyzed and Metallocene Polyethylenes 231
    Paula Wood-Adams and John M. Dealy
The Influence of Sequence Length Distribution on the Linear Crystal Growth
of Ethylene-Octene Copolymers 239
    John E. Wagner, Samir Abu-Iqyas, Kenneth Monar and Paul J. Phillips
Crystallization and Microstructure of Ziegler-Natta and Metallocene Based
Isotactic Polypropylenes: Simulation and Experiment 247
    Y. Churdpunt and A. I. Isayev
Crystallization of Ethylene-Octene Copolymers at High Cooling Rates 257
    Paul J. Phillips and John Wagner
Kinetics of Non-Isothermal Crystallization Syndiotactic Polypropylene:
Avrami, Ozawa, and Kissinger Approaches 263
    Pitt Supaphol and Joseph E. Spruiell
High Temperature Flexible Polyolefins a User's Perspective 275
    Atul Khare, Samuel Y Ding, Michael T. K. Ling, and L. Woo

Markets

An Outlook for Metallocene and Single Site Catalyst Technology into
the 21st Century   283
    Kurt W. Swogger
World Olefins Markets: Issues Influencing Availability and Pricing   293
    Robin G. Harvan
World Polypropylene Demand   299
    Russ DeLuca
PVC Markets, Today and in 2020   303
    Donald Goodman
Index   307

Preface

The revolution in polyolefin technology started during the last decade of the 20th century continues today at an ever-faster pace. We are talking about the metallocene and single site catalyst based homo and copolyolefins (mPOs) with their well-defined structures and closely controlled molecular architectures. Their capabilities to compete, on the basis of their tailored properties against traditional polymers such as ABS, PVC, thermoplastic elastomers (TPEs) and others combined with a favorable cost structure constitutes a fertile resource for improved materials.

Since the publication last year of "Metallocene Catalyzed Polymers", where the advances in catalysis and polymerization techniques where discussed in detail together with the polymer properties, processing and the initial markets, we have witnessed a tremendous growth in the commercial development of these novel materials. They are more and more targeted on an improved cost/performance basis at replacing or complementing the traditional materials. We have also witnessed some inroads of the novel materials in other areas not contemplated initially such as elastomers, adhesives and engineering thermoplastics.

One of the inevitable but expected aspects of the new product/process development in a free market economy is the protection of intellectual property. In many cases this leads to long drawn and expensive litigation between major or minor players with adverse effects on timely commercial development. Therefore the recently announced (see C&E News June 21st, 1999) agreement between Dow Chemical Co., Exxon and Univation Technologies (a licensing joint venture between Exxon and Union Carbide Corp.) to settle rather than litigate their patent rights constitutes certainly a step in the right direction of an increased competitive market growth for metallocene polyolefins. This is important since besides the two major players, Dow Chemical Co. and Exxon, many more companies have announces their commercial efforts for metallocene polyolefins. (BASF, DSM, TOTALFINA, MITSUI, BOREALIS, EQUISTAR, Union Carbide, etc)

The present book is based on 40 papers presented at the ANTEC 98 and 99 and other conferences. As a general observation, the center of gravity has shifted from polymer manufacturing to application development. As a consequence there are only two catalysis related papers in the chapter on **Catalytic Systems**. They deal with the polymerization results using supported single site catalysts (for mPEs), as well as with the detrimental effect of the residual triethyl aluminum on the molecular weight of elastomeric PP.

The bulk of the book describes, in the chapter of **Processing**, the development work in adapting these new, now commercial materials into specific applications, using appropriate processing methods and techniques.

In some cases metallocene polyolefins can be used in applications directly. For instance the second-generation metallocene PEs (EZP-mPE) presented by Univation Technologies exhibit controlled MWD (compared with the first generation that was very narrow, and thus harder to process), compositional distribution, long chain branching and distribution. These structural properties lead to a much tougher film in comparison with the typical blends of Ziegler-Natta ZN-LL/LDPEs. Consequently thinner films are possible, this downgauging resulting in cost savings. Additionally, the heat scalability of mPEs is also superior.

In many other cases one unique material cannot provide the entire range of desired properties. In this case blends are preferred, and this is no different for mPOs. Blends can be formulated to target existing applications, while taking advantage of the properties afforded to the blend by the mPOs. Whenever the use of blends is necessary one has to tackle the issues of miscibility and compatibility. It's worth directing the reader's attention at a new concept in polymer blend, presented by a group from Exxon, namely the spherulite boundary strengthening. This is achieved by a copolymer that migrates to the spherulite boundary and cocrystallizes creating additional interspherulitic links. The strengthening agent studied was an isotacticPP-atacticPP made by single site catalysis, containing additionally sodium benzoate as nucleating agent. Another interesting blend application presented by Dow Chemical Co. workers uses an ethylene-styrene interpolymer made using INSITE technology. This product could be blended with either atactic PS, or PP, giving unique properties. Uses as moldings, films and foams are contemplated.

All major processing techniques have been impacted by the novel mPOs: extrusion, film blowing, injection molding, foam production, fiber spinning and composite manufacturing. Efforts in process parameter optimization, using the new mPOs and their blends and reported in several of the papers, are included in the **Processing** chapter. They encompass film applications for fresh food packaging, heat sealing, coextrusion, PP films and applications, nitrogen autoclave foaming and others.

Separately we grouped the papers studying **Structure-Property** relationships, including rheology, instrumental analysis, study of crystallization, crystal growth kinetics.

In a final chapter we grouped the papers dealing with polymer **Markets**, market outlook for mPOs, the availability and pricing of monomers, projected world demand for PP as well as a long term look at PVC markets in the next two decades. The PVC study is included because of the expected future market pressure on the flexible PVC applications from the metallocene polyolefins.

We do expect indeed to see in the future a continuation of the metallocene polyolefin revolution. Today's technology appears to be only in a "transition state". As competition in the market place will accelerate and, in turn, as the market place will exert more stringent performance demands, novel, higher productivity catalytic systems will temporarily win the popularity contest. We expect to see the introduction of some products containing functional groups, we also expect competition from catalytic systems based on Ni, Fe Pd, rather than Zr, Ti, Hf.

George M. Benedikt
The BFGoodrich Company
Brecksville, Ohio
August, 1999

Single-Site Supported Catalysts for Ethylene

A. Muñoz-Escalona, L. Mendez, B. Peña, P. Lafuente, J Sancho, W. Michiels,
G. Hidalgo and Mª. Fca. Martinez-Nuñez

Repsol, S.A. c/Embajadores, 183, 24085 Madrid, Spain

INTRODUCTION

Single-site catalysts emerged in the polymers field with the discovery of metallocene catalysts by Sinn and Kaminsky[1] based on early transition metals (Zr, Ti, Hf). Soon after their appearance in 1980 their advantages over the conventional multi-site Ziegler-Natta and chromium catalysts were recognized. Thus, they are high active catalysts exhibiting an exceptional ability to polymerize olefin monomers, producing extremely uniform polymers and copolymers of narrow molecular weight distribution (polydispersity of about 2) and narrow chemical compositional distribution, controlling at same time the resulting polymer chain architectures. The evolution of single-site coordinative catalysis continue with the appearance in 1995 of new Ni and Pd diimine complexes capable of producing hyper-branched and functionalized polyolefins[2] and more recently in 1998 with the synthesis of highly active catalysts for ethylene polymerization based on substituted tridented pyridine bis-imine ligands attached to Fe and Co metals.[3]

To prepare supported single-site catalysts different methods can be employed.[4,5] Among them, the most investigated methods are based on direct adsorption over inorganic supports (Method 1) or on previously passivated supports by treatment with methylaluminoxane, MAO, (Method II). A different method, based on the synthesis of metallocenes with functional chemical groups and supporting them by reaction with the chemical groups of the carriers, has been recently published.[5] The reactivity between the functional groups of the carriers and the metallocene are carefully controlled so that the reaction is very selective, allowing a uniform distribution of the active sites across the catalysts particles.

We present results on supported metallocene single-site catalysts using the above mentioned methods as well as results obtained with unsupported and supported single-site catalysts based on Ni and Fe metals using method II.

EXPERIMENTAL

All solvents and air-sensitive compounds were handled under nitrogen atmosphere using Schlenk techniques.

Commercial carriers: silica (Crossfield, UK and PQ, USA) and MAO modified silica were employed as received. The preparation of supported catalysts was carried out by suspending the carrier in toluene and adding the organometallic compound. The resulting supported catalyst was carefully washed with additional toluene, filtered and dried under vacuum to give a free flowing powder. The metal contents were measured by inductively coupled plasma (JCP) spectroscopy.

Slurry polymerizations were carried out in batch using a 1,2 liter glass reactor autoclave filled with 600 ml of n-heptane, 4 bar ethylene pressure and at different reaction temperatures. The catalysts concentration in the reactor was in the range of 0,0056-0,0017 mmol depending of the metal used. MAO and triisobutylaluminium (TIBA) were used as co-catalysts at a M/Al ratio of 700.

The polymers were characterized by GPC, TREF and NMR ^{13}C using well established techniques. Finally, the morphologies of catalysts and polymers particles were observed under scanning electron microscope (SEM) coupled with an EDX detector for metals analysis. In order to examine the distribution of the metals across the catalyst particles or the inner morphology of the polymer particles, they were previously sectioned.

RESULTS AND DISCUSSION

SUPPORTED METALLOCENE SINGLE-SITE CATALYSTS.

When Method I is used to prepare SiO_2 supported metallocene catalysts, different type of complexes covalently and ionically attached to the silica are formed on its surface.[5] Using thermogravimetric analysis coupled to a mass spectrometer it could be also observed that this reaction is accompanied by a significant decomposition of the metallocene as some amount of free ligands could be detected. As a consequence, only a small fraction of the total supported metallocene (approximately 1%) is active for the ethylene polymerization. Therefore, it is not surprising that the activities found for catalysts are one to two orders of magnitude lower compared to the homogeneous analogs. Finally, if MAO is used as co-catalyst leaching of the metallocene was found to occur so that the polymerization takes place not only on the catalyst particle but also in solution. As result, no replication phenomena was found and same reactor fouling or sheeting was observed.

By using Method II for supporting the metallocene, the detrimental effect of the Bronsted and Lewis groups of the silica can be avoided. The active catalytic species are directly formed so that no additional MAO is needed for activation and only TIBA as scavenger

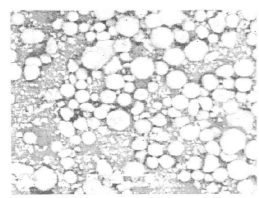

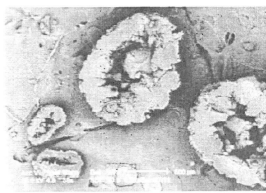

Figure 1. Morphology of MAO modified (MAO/SiO₂) containing 23% Al.

Figure 2. Cross-sectioned PE particles produced with supported metallocene on MAO/SiO₂.

for impurities is usually needed. Furthermore, the metallocene complex is bound to the support only by loosely ionic interaction so that it can float and move over the supported layers of the MAO resembling the homogeneous catalyst. Therefore, catalysts with higher activities than those obtained in Method I can be obtained. Regarding the control of the polymer morphology by the catalyst, one important factor to be considered is the morphology of the starting support MAO/SiO₂. In Figure 1 the morphology of a MAO modified SiO₂ containing 23% Al is shown. It can be observed that samples are composed by big spherical grains together with small irregular particles detached from the former. By EDX analysis under the SEM it could be shown that the small particles are debris detached of the outer layer of practically pure MAO covering the silica. The small particles can be almost removed by washing them out with toluene. Furthermore, the content of MAO across the support is not uniform distributed decreasing its concentration from the surface to the inside of the big silica particles. After the absorption of MAO on the silica catalysts were prepared by addition of the metallocene compound. The immobilization of the metallocene proceeds trough extraction of one methide (or Cl) ligand by the MAO giving rise to the formation of cation-like metallocenium-active species. Due to the uneven distribution of MAO in the modified MAO/SiO₂ a high concentration of active species are produced in the outer layers of the catalyst decreasing gradually toward the center of the particles. Therefore, a non-uniform fragmentation of catalyst particles occurs during polymerization. The rapidly expanding outer layers are not followed by the less active growing inner part resulting in hollow poorly filled polymer particles. Sectioned polymer particles produced with this type of catalysts can be seen in Figure 2. Furthermore, due to the very fast fragmentation of the outer layers of the catalysts, considerable amount of fine polymer particles is also produced. Finally, due to the

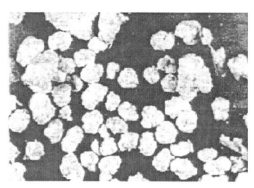

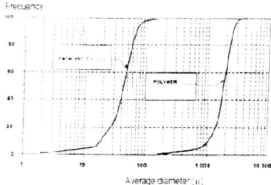

Figure 3a. Morphology of PE produced with supported functionalized metallocene on SiO₂.

Figure 3b. Replication phenomena obtained with supported functionalized metallocene on SiO₂.

mobility of the active catalytic species migration to the reactor walls or reaction medium take place leading to reactor fouling or sheeting, specially at high polymerization temperatures.

The above mentioned disadvantages can be overcome by synthesizing metallocene with functional chemical groups and reacting them with the carrier according to the reaction scheme below.

L= Cp, Indenyl, Fluorenyl; X= halogene

The active species are attached to the silica by covalent bonds so that no leaching can take place avoiding reactor fouling. The mobility of the active catalytic species are controlled by the number of functional groups of the metallocene reacting with the carrier and the length of the spacers placed between the carrier and the metallocene. Furthermore, using porous carriers an even distribution of the anchored metallocene across the particle is obtained, leading to a uniform fragmentation of the catalyst and to a good replication phenomena, as shown in Figure 3.

SUPPORTED Ni BASED SINGLE-SITE CATALYSTS

The aryl-bis-diimine complexes of Ni were synthesized after Brookhart's procedure.[2] The soluble catalysts and also heterogenized by supporting them on MAO modified SiO₂ (Method II) with 23% of Al were used for ethylene polymerization (see structure on the next page).

R = Me

/ MAO

The activities, molecular weights (MW), and polydispersities of the polymers produced by supported and unsupported catalysts at different temperatures are given in Table 1. Like for metallocene catalysts, polymerization activities for supported catalysts are one order of magnitude lower compared to the homogeneous analogs. Also the MWs of the polymers follow the same behavior (e.g., heterogeneous catalysts give higher MWs compared the to homogeneous soluble catalyst). With both catalytic systems, activities and MWs decrease with polymerization temperature. Furthermore, due to the migratory insertion mechanism highly branched PEs are obtained. The most common branches found are methyl followed by hexyl type. Heterogeneous catalysts give PEs with lower branches content compared to the homogeneous catalysts. With both catalytic systems the number of branches increases with temperature.

Regarding the polymer morphology, polymers produced with supported Ni catalysts exhibit similar morphologies as supported metallocene catalysts using the same support and method of preparation. Thus, Figure 4a shows a general view of the morphology of polyethylene fluff obtained at 45°C polymerization temperature. Well formed polymer particles can be seen together with fine particles, which are produced when very active catalysts particles blow up due to their very fast expansion at the very early polymerization stages. In Figure 4b, sectioned hollow particles are similarly shown.

Table 1. Characterization of PEs produced with homogeneous and heterogeneous Ni-based catalysts

Catalyst	Temp °C	A*	M_w	M_w/M_n	B**	Me	Et	Bu	Pen	Hex
Homogeneous	45	13.9	135300	2.5	34.8	22.2	2.1	2.7	1.7	5.1
Homogeneous	70	3.07	33500	1.9	79.5	53.1	5.8	3.2	2.8	13.5
Homogeneous	80	2.47	-	-	94.2	57.0	7.2	5.6	4.4	16.2
Homogeneous	90	1.88	-	-	101.7	58.0	9.6	5.6	3.8	19.5
Homogeneous	100	1.10	-	-	-	-	-	-	-	-
Heterogeneous	45	1.39	242900	2.8	18.4	15.1	3.3	0.0	0.0	0.0
Heterogeneous	60	0.71	155200	2.8	32.7	25.3	3.1	1.2	0.0	3.1
Heterogeneous	70	0.54	95800	2.8	48.1	36.0	3.7	1.5	1.1	5.1

* Activity x10^{-6} (g PE/mol bar h), **Total branches

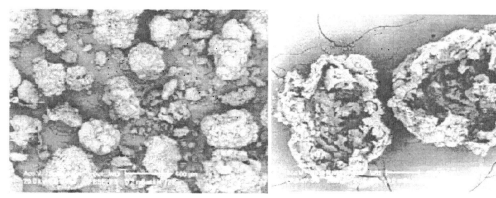

Figure 4a. Morphology of PE obtained with Ni-based catalyst   Figure 4b. Cross-section of PE particles.
supported on MAO/SiO$_2$.

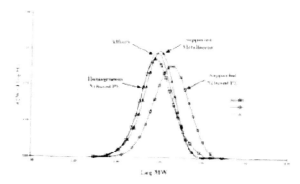

Figure 5. GPC curves of affinity and PEs obtained with supported
metallocene and homogeneous and supported Ni-based catalysts.

The question is whether catalysts can be considered as single-site or not. To answer this question GPC and TREF analysis of the polymers were carried out. Narrow molecular weight distribution (MWD) of polymers produced with supported metallocene as well as with Ni based catalysts (supported and unsupported) are presented in Figure 5. For comparison, PE Affinity from Dow produced with metallocene in solution has been also included. More information can be obtained from TREF analysis of copolymers as given in Figure 6. It can be seen that the different (co)polymers exhibit very narrow chemical composition distributions. Thus, the 1-hexene metallocene copolymer having 12.1 SCB/100°C present a narrow peak in the TREF curve. The high branched 1-octane copolymer with 23 SBC/100°C (Affinity plastomer from Dow) present a small shoulder on the TREF trace at about 92°C elution temperature. The Ni based hyper-branched PE (with 29/100°C SCB content) presents also a narrow peak at 81.1°C elution temperature but a more visible peak at about 94°C. The differences observed for both PE types (Affinity and Ni based PE) could be due to the difference in the amount and type of SCB: hexyl in case of the Affinity and a mixture of ethyl, buthyl, hexyl and predominantly methyl for the more branched Ni based PE. The similarity between the Ni hyper-branched polyethylene and the Affinity plastomers produced by

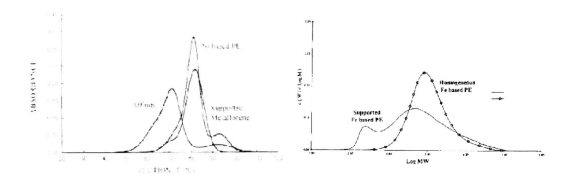

Figure 6. TREF curves of affinity. Ni-based PE (see Table 3)   Figure 7. GPC curves of PE produced with Fe-based catalysts.
and supported metallocene based PE.

Table 2. PEs produced with Fe-based catalyst

Catalyst	Temperature, $^{\circ}$C	Activity, g PE/mol bar h	M_w	M_w/M_n
Homogeneous	45	16	255900	4.1
Homogeneous	60	16	133100	11.1
Homogeneous	70	23	157900	10.7
Homogeneous	80	0.0	-	-
Heterogeneous	45	4.5	631800	5.7
Heterogeneous	60	5.2	413200	6.9
Heterogeneous	80	1.9	327600	27.1
Heterogeneous	90	1.5	245000	24.5

Table 3. Physical chemical characterization and mechanical properties of Affinity and Ni-based PE

PE	M_w	M_w/M_n	B*	Me	Hex	LCB	Density g/ml	Crystal %	YM** MPa
Ni-based PE	118000	2.9	29	21	5	0.94	0.8965	27	63
Affinity	105000	2.4	23	-	23	0.41	0.9010	18	68

* Total branches/1000C, ** Young modulus

copolymerization of ethylene with 1-octene can be found in the density, % of crystallinity and mechanical properties, as shown in Table 3 (see also reference 5). These results indicate that with Ni catalysts plastomer Affinity type PEs could be produced.

SUPPORTED Fe BASED SINGLE-SITE CATALYSTS.

Finally, the following Fe based single-site catalyst was synthesized after Bookhart *et al.*[3] and used supported and unsupported for ethylene polymerization (see structure on this page).

The Method II used for supporting this catalyst is the same as the one used for Ni based catalyst. However, the main difference between both supported catalysts is that the Fe based catalyst does not polymerize ethylene when TIBA is used as co-catalyst. Therefore, to produce an active catalyst, additional MAO is needed in the reaction medium. The activities are lower compared to the Ni based catalysts but the MWs are higher. The supported catalysts show activities at temperature higher that 80°C, while the homogeneous catalysts not. For homogeneous and supported catalysts the polydispersities are too high for a single-site catalyst. The MWDs are very broad and even bimodal as shown in Figure 7. Therefore, more research is needed for producing Fe supported single-site catalyst. For instance by exploring different type of systems for catalyst activation.

REFERENCES

1　H. Sinn and W. Kammisky. *Adv. Organometallic Chem.*, 1980, **18**, 99.
2　L. K. Johnson, Ch. Killian and M. Brookhart. *J. Am. Chem. Soc.*, 1995, **117**, 6414.
3　B. L. Small, M. Brookhart and A. M. A. Bennet. *J. Chem. Soc.*, 1998, **120**, 4049.
4　M. R. Ribeiro, A. Deffleux and M. F. Portela. *Ind. Eng. Chem. Res.*, 1997, **36**, 1224.
5　A. Munoz-Escalona, L. Mendez, J. Sancho, P. Lafuente, B. Peaea, W. Michels, G. Hidalgo and M. F. Martinez-Nunez. To be published in Metalorganic Catalysts for Synthesis and Polymerization. Springer Verlag, Berlin (Germany), 1998.

Synthesis of High-Molecular-Weight Elastomeric Polypropylene With Half-Titanocene/MAO Catalysts

Oing Wu, Benheng Xie, Meiran Xie, Rui Chen
Institute of Polymer Science, Zhongshan University, Guangzhou 510275, China

INTRODUCTION

The stereochemistry and molecular weight of polyolefins strongly influence their physical and mechanical properties. It is well known that isotactic and syndiotactic polypropylenes consisting of a regular arrangement of the stereocenters are crystalline thermoplastics with high melting points. Atactic polypropylene is an oil, a sticky soft solid, or an amorphous elastomer, depending upon its molecular weight being low, middle, or high.

Natta[1,2] was the first to produce rubbery polypropylene by titanium chloride catalyst in the 1950s and interpreted the elastomeric properties of this material as a result of stereoblock structure composed of crystallizable isotactic stereosequences and amorphous atactic sequences. Elastomeric stereoblock polypropylenes with low isotactic content and crystallinity and a high molecular weight ether-soluble fraction can be obtained by polymerization using alumina-supported tetraalkyl[3] or bisarene[4] group IVB catalysts. Significant efforts have been directed to study the polymerization of propylene catalyzed by homogeneous metallocene-aluminoxane catalyst systems since the 1980s. These results indicate that higher steric controllability with the cyclopentadienyl ligand allows the synthesis of many metallocenes with different symmetries which give rise to different stereoregular arrangements of polymers. A number of non-symmetric ethylidene-, propylidene- and dimethylsilylene-bridged metallocene complexes of titanium, zirconium, and hafnium have been employed in preparing the thermoplastic-elastomeric polypropylenes.[5-7] A mechanism of polymerization was postulated invoking an active species having two interconverting propagating states. The polymerization occurs alternately on the aspecific and stereospecific sites to form alternating stereoregular/crystallizable and stereoirregular/amorphous segments. A unbridged metallocene catalyst composed of bis(2-phenylindenyl)zirconium dichloride and MAO was designed to isomerize between chiral rac-like and achiral meso-like geometries by rotation of the indenyl ligands about the metal-ligand bond axis in order to obtain atactic-isotactic stereoblock elastomeric polypropylene.[8] Recently, a novel, atactic polypropylene with high molecular weight has been prepared with C_{2v} symmetric dimethylsilanediylbis(9-fluorenyl)zirconium dichloride and dimethyl/MAO catalyst systems.[9]

On the other hand, half-titanocenes have been reported to be efficient catalyst precursors for syndiospecific polymerization of styrene as well as catalyzed polymerizations of substituted styrene, conjugated diene, ethylene and their copolymerization. Very little work on propylene polymerization by monocyclopentadienyl catalysts has been reported in open literature.[10-12] Recently, Bochmann et al.[13] reported a catalyst system generated from 1:1 mixtures of Cp*TiMe$_3$ and B(C$_6$F$_5$)$_3$ being highly active for propylene polymerization to give high-molecular-weight atactic and elastomeric polypropylene. As revealed by GPC analysis of the produced polymer, a significant proportion of the titanium centers acts as living propylene polymerization catalyst.

In this paper, we report a new result from aspecific polymerizations of propylene promoted by half-titanocene/MAO through controlling the amount of the residual TMA in MAO. The dependencies of molecular weight of the atactic polypropylene on the structure of titanocene and polymerization conditions are also reported.

EXPERIMENTAL

MATERIALS

Trimethyl aluminum (TMA) was commercially available and used without further purification. Toluene, benzene and hexane were distilled from sodium under nitrogen just before use. Half-titanocene compounds, CpTi(OBz)$_3$, CP*TiCl$_3$, Cp*TiBz$_3$, Cp*Ti(O allyl)$_3$ and Cp*Ti(OBz)$_3$ (Cp=η-cyclopentadienyl, Cp*= η-pentamethylcyclopentadienyl, Bz = benzyl), were synthesized by the similar procedures as in literature.[14,15] MAOs were prepared as follows: 200 mL of TMA solution (3.1 M in toluene) was added dropwise into a flask with appropriate amount of ground Al$_2$(SO$_4$)$_3$·18H$_2$O in toluene at 0°C. The mixture was gradually heated to 60°C and stirred for 24 h, and was then filtered. The filtrate was concentrated under reduced pressure to a white solid. Residual TMA content in the MAO was determined by pyridine titration. The MAOs prepared with initial [H$_2$O]/[TMA] molar ratios of 1.0, 1.3 and 1.8 contain the residual TMA 28.0%, 22.9% and 15.4%, respectively, and are referred to MAO1, MAO2 and MAO3. MAO2 was redissolved in toluene and then the solvent was vaporized at 60°C in vacuo, to give MAO2'.

POLYMERIZATION PROCEDURE

Polymerizations were carried out in a 100 mL glass flask equipped with a magnetic stirrer. Twenty mL of toluene saturated with the monomer and desired amount of MAO and titanocene compound were introduced in this order. The gaseous monomer was fed on demand. The total pressure was maintained at 130 kPa throughout the course of the polymerization. The polymerizations were stopped after 1 h and the polymers were precipitated by addition of acidified alcohol. The resulting polymers were washed with alcohol and dried in vacuo to constant weight.

CHARACTERIZATION

Molecular weight of polymer was obtained from intrinsic viscosities[16] measured at 135°C in tetrahydronaphthalene. The molecular weight distributions were measured on a Waters-208 LC/GPC with chloroform as solvent at 40°C. Monodisperse polystyrene was used as a standard. ^{13}C NMR spectra of the polymers in o-dichlorobenzene were recorded on a Bruker-AM400 at 120°C. DSC scans were obtained on a PE DSC-7c at a scanning rate of 10°C/min from -40 - 200°C.

RESULTS AND DISCUSSION

EFFECT OF RESIDUAL TMA

MAO is the ubiquitous co-catalyst used in metallocene catalyzed olefin polymerization. It always contains some amount of unreacted TMA that cannot be removed easily by vacuum distillation. The residual TMA can be a factor influencing the catalytic activity of metallocenes and molecular weight of polyolefins. Four MAOs with various amounts of residual TMA were tested as cocatalyst in the propylene polymerization with catalyst precursors CpTi(OBz)$_3$ and Cp*Ti(OBz)$_3$. The activity of the catalyst is very sensitive to the TMA content in MAO, as showed in Table 1. Using MAOs with residual TMA content more than 20 mol% (MAO1 and MAO2), the catalysts show considerably poor activities for propylene polymerization. High catalytic activity is achieved only with MAO3 which contains 15.4 mol% of TMA. Reducing the amount of residual TMA from 22.9 mol% (MAO2) to 17.4 mol% (MAO2') by redissolving MAO2 in toluene and then vaporizing the solvent in vacuo at 60°C causes a dramatic increase in catalytic activity. On the contrary, addition of external TMA into MAO3-containing system, which originally gives high activity in propylene polymerization, deactivates the catalyst.

To give a better insight into the effect of the TMA in MAO on the polymerization, oxidation state distribution of the Ti species was measured by redox titration[17] under the similar conditions as the polymerizations. The values of titanium oxidation states from the reaction

Table 1. Polymerization of propylene using half-titanocene activated with various MAOs[a]

| Titanocene | MAO preparation | | | External TMA mM | Oxidation states[b], mol% | | | Activity kg PP/mol Ti h | M_w^c x10^{-4} | T_g °C |
	No.	H_2O /Al	TMA %		Ti^{4+}	Ti^{3+}	Ti^{2+}			
CpTi(OBz)₃	MAO1	1.0	28.0	0	12.1	83.3	4.7	trace		
	MAO2	1.3	22.9	0	15.4	80.9	3.7	2.20		
	MAO3	1.8	15.4	0	75.3	23.6	1.1	87.7	5.3	-8.1
	MAO3	1.8	15.4	16.7	29.4	67.4	3.2	trace		
	MAO2'	d	17.4	0	73.7	25.7	0.6	58.9	4.4	-9.2
Cp*Ti(OBz)₃	MAO2	1.3	22.9	0	25.1	60.0	14.9	trace		
	MAO2'	d	17.4					62.7	31.5	-7.3
	MAO3	1.8	15.4	0	81.3	17.3	1.1	72.0	41.9	-7.8[a]

[a]Polymerization conditions: [Ti]=1 mM; [MAO]=170 mM; toluene 20 mL; 40°C, 1 h. [b]Ti oxidation states were measured by redox titration. [c]Molecular weight from intrinsic viscosity measurement. [d]MAO2' was obtained by re-dissolving MAO2 in toluene and then vaporizing the solvent at 60°C in vacuo.

of the titanocenes with MAO are given in Table 1. Large fractions of titanium are in the reduced oxidation states ($[Ti^{3+}]$ = 80.9-83.3% and $[Ti^{2+}]$ = 3.7-4.7% for CpTi(OBz)₃, 60% and 14.9% for Cp*Ti(OBz)₃) when the titanocenes react with MAO1 or MAO2 having high TMA content, whereas the reactions with MAO3 and MAO2' containing less TMA give mostly Ti^{4+} species ($[Ti^{4+}]$ = 73.7-75.3% for CpTi(OBz)₃, 81.3% for Cp*Ti(OBz)₃). With the addition of a small amount of external TMA into the catalyst system with MAO3 originally containing less residual TMA, $[Ti^{3+}]$ increases remarkably from 23.6% to 67.4% and $[Ti^{2+}]$ increases from 1.1% to 3.2%, and the catalytic activity drops down correspondingly.

The results indicate that the catalyst activity can be correlated with titanium valence states. Titanium in the catalyst systems with MAO containing less residual TMA can remain mostly in a tetravalent state, and promotes propylene polymerization efficiently. On the other hand, a high TMA content in MAO or addition of external TMA favors the reduction of titanium to the lower valent states which do not catalyze the polymerization of propylene. Therefore, we propose that the active species for the propylene polymerization in the half-titanocene/MAO systems are in tetravalent state, and the Ti species in the reduced valences are inactive for propylene polymerization, but active for styrene syndiotactic polymerization.[14,17,18]

Table 2. Polymerization of propylene with different half-titanocenes activated with MAO3

Titanocene	[MAO3]/[Ti] mol/mol	Activity kg PP/mol Ti h	M_w x 10^{-4}	Ether-soluble fraction, %
CpTi(OBz)$_3$	330	84.9	4.7	99
Cp*TiCl$_3$	330	37.6	12.9	98
Cp*Ti(OBz)$_3$	330	98.0	42.6	99
Cp*Ti(O allyl)$_3$	330	134	34.5	98
Cp*TiBz$_3$	230	112	34.3	nd.

Cp=cyclopentadienyl; Cp*=pentamethylcyclopentadienyl; [Ti]=0.5 mM; toluene 20 mL, T=40°C, t=1 h

DEPENDENCE OF THE PRODUCT MOLECULAR WEIGHT

Five half-titanocenes, CpTi(OBz)$_3$, Cp*TiCl$_3$, Cp*TiBz$_3$, Cp*Ti(OBz)$_3$ and Cp*Ti(O allyl)$_3$, were used as the catalyst precursor in the propylene polymerization with co-catalyst MAO3. As shown in Table 2, the titanocene without substitution for H at the Cp ring gives atactic polypropylene of modest molecular weight (Mw = 4.4×10^4), while the related one having pentamethyl substitution at Cp ring gives high molecular weight atactic polypropylene (Mw=41.9×10^4) under the same conditions, increasing molecular weight by one order of magnitude. On the other hand, esterified or alkylated titanocenes with appropriate -OR and -R ligands produce higher molecular weight polypropylene than the corresponding halide.

The most common cause for molecular weight depression in metallocene-catalyzed polymerizations is β-hydrogen transfer. The higher molecular weight of the atactic polypropylenes obtained by catalyst systems Cp*Ti(OR)$_3$/MAO and Cp*TiR$_3$/MAO can be mainly due to an increase in electron density at the metal center and also at β-carbon of the growing polymer chain which is caused by substitutions of more electron-releasing Cp* and -OR or -R ligands for Cp and chlorine, respectively. It makes the thermodynamic driving force for β-H transfer diminishing. In addition, bulk of the Cp* and -OR or -R ligands has certain steric effect on β-hydrogen transfer by increasing the energy of the transition state for which it is necessary to rotate the growing polymer chain about the C_α—C_β bond.

The polymers produced by the Cp*-bearing titanocene catalysts at ambient pressure and temperatures ≤ 60°C have molecular weight higher than 20×10^4 with narrow polydispersities ranging from 1.5 to 2.0. As shown in Table 3, the molecular weight of the polymers is also sensitive to polymerization temperature and increases with reducing the temperature. When the polymerizations were carried out at 0°C in toluene, Mw values of the produced polypropylenes reach 103×10^4 for Cp*Ti(OBz)$_3$ system, 69×10^4 for Cp*Ti(O allyl)$_3$ system,

Table 3. Polymerization of propylene at different temperatures

Titanocene	T_p °C	Activity kg PP/mol Ti h	$M_w \times 10^{-4}$	Ether-soluble fraction, %
Cp*Ti(OBz)₃	0	15.0	103.7	96
	20	27.9	75.7	97
	30	55.6	44.5	nd.
	40	72.0	41.9	99
	50	36.6	22.0	nd.
	60	8.8	18.7	99
Cp*Ti(O allyl)₃	0	112	69.0	96
	20	110	56.5	99
	40	113	27.3	99
	60	51.0	18.0	nd.
Cp*TiBz₃	-17	21.8	83.9	nd.
	0	116	55.1	nd.
	20	112	34.3	nd.
	40	40.0	28.3	nd.
	60	17.6	21.6	nd.

Polymerization conditions: [Ti]=1 mM; [MAO3]/[Ti]=170; toluene 20 mL; t=1 h

Table 4. Effect of temperature on propylene polymerization in hexane with Cp*Ti(OBz)₃/MAO3 system

T_p, °C	Activity, kg PP/mol Ti h	$M_w \times 10^{-4}$	Ether-soluble fraction, %
0	27.1	85.2	93
20	51.4	51.5	97
30	68.3	42.8	nd.
40	89.3	35.6	98
50	68.7	20.3	nd.
60	25.7	16.1	99

Polymerization conditions: hexane 20 mL; T=40°C; other conditions the same as in Table 3

and 55.1×10^4 for Cp*TiBz₃ system. The half-titanocene catalyst systems are also suitable for producing high-molecular-weight atactic polypropylene in saturated hydrocarbon solvent. Similar result was obtained from the polymerizations in hexane (see Table 4).

All the polypropylenes obtained with the Cp*-bearing titanocene catalysts are amorphous, elastomeric materials and soluble in ether and hydrocarbon solvents. The glass transitions of the materials appear approximately at -8°C by DSC measurement. ^{13}C NMR analyses confirm the half-titanocene/MAO catalysts afford atactic and, to a certain extent, regio-irregular polypropylenes.

ACKNOWLEDGMENT

Financial support of this research from the National Natural Science Foundation of China and the Science Foundation of Guangdong Province are gratefully acknowledged.

REFERENCES

1   G. Natta, G. Mazzanti, G. Cresci, G. Moraglio, *Chim. Ind, Milan*, **39**, 275 (1957).
2   G. Natta, *J. Polym. Sci.*, **34**, 531 (1959).
3   J. W. Collette, C. W. Tullock, R. N. MacDonald, W. H. Buck, A. C. L. Su, J. R. Harrell, R. Mulhaupt, B. C. Anderson, *Macromolecules*, **22**, 3851 (1989).
4   C. W. Tullock, F. N. Tebbe, R. Mulhaupt, D. W. Ovenall, R. A. Setterquist, S. D. Ittel, *J. Polym. Sci., Part A: Polym. Chem.*, **27**, 3063 (1989).
5   D. T. Mallin, M. D. Rausch, Y. G. Lin, S. Dong, J. C. W. Chien, *J. Am. Chem. Soc.*, **112**, 2030 (1990).
6   G. H. Llinas, S. H. Dong, D. T. Mallin, M. D. Rausch, Y. G. Lin, H. H. Winter, J. C. W. Chien, *Macromolecules*, **25**, 1242 (1992).
7   W. J. Gauthier, J. F. Corrigan, N. J. Taylor, S. Collins, *Macromolecules*, **28**, 3771 (1995).
8   G. W. Coates, R. M. Waymouth, *Science*, **267**, 217 (1995).
9   L. Resconi, R. L. Jones, A. L. Rheingold, G. P. A. Yap, *Organometallics*, **15**, 998 (1996).
10 K. Soga, D. H. Lee, *Makromol. Chem.*, **193**, 1687 (1992).
11 C. Pellecchia, A. Proto, P. Longo, A. Zambelli, *Makromol. Chem., Rapid Commun.*, **13**, 277 (1992).
12 Q. Wu, Z. Ye, Q.-H. Gao, S.-A. Lin, *J. Polym. Sci., Part A: Polym. Chem.*, **36**, 2051 (1998).
13 J. Sassmannshausen, M. Bochmann, J. Roesch, D. Lilge, *J. Organomet. Chem.*, **548**, 23 (1997).
14 Q. Wu, Z. Ye, S.-A. Lin, *Macromol. Chem. Phys.*, **198**, 1823 (1997).
15 M. Mena, P. Royo, R. Serrano, M. A. Pellinghell, A. Tiripicchio, Organometallics, 8, 476 (1989).
16 D. Pearson, L. Fetters, L. Younghouse, J. Mays, *Macromolecules*, **21**, 479 (1988).
17 J. C. W. Chien, Z. Salajha, S. Dong, *Macromolecules*, **25**, 3199 (1992)
18 P. Longo, A. Proto, A. Zambelli, *Macromol. Chem. Phys.*, **196**, 3015 (1995).

Is Metallocene Polyethylene Blend with HDPE More Compatible Than with PP?

D. Rana, K. Cho, B. H. Lee and S. Choe
Department of Chemical Engineering, Inha University, Inchon 402-751, Korea
Taedok Institute of Technology, SK Corporation, Taejon 305-370, Korea

INTRODUCTION

Polyolefins are the prime polymers in the industrial field. A vast amount of blends in linear low density polyethylene (LLDPE) with conventional polyolefins have been commercially used in the agricultural application and packaging industry as a form of extrusion blown films. LLDPE contains generally 1-butene or 1-hexene or 1-octene comonomer controlled by Ziegler-Natta catalyst. Recently LLDPEs with uniformly distributed high percentage of comonomer are made by metallocene catalysts. Though many research works[1-5] have already been done regarding LLDPE made by Ziegler-Natta catalyst, it still needs to study recently developed metallocene polyethylene (MCPE). The interfacial tension between MCPE and polypropylene (PP) was found to be very low which indicates submicron dispersions of polyolefins were accomplished even with high melt index of PP.[6] Again MCPE exhibits sharp relaxation at sub-ambient temperature.[7] High and low molecular weight metallocene catalyzed high density polyethylene (HDPE) formed a miscible blend.[8] Previously this laboratory had a project regarding miscibility and processability of LLDPE with other conventional polyethylene.[9-15] It is our interest to examine the thermal, mechanical and viscoelastic behaviors of MCPE blended with conventional polyolefins, e.g., HDPE and PP. In this article, we present evidence, first to our knowledge, that both blends are immiscible and the extent of compatibility of the HDPE-MCPE blends is larger than the PP-MCPE blends by both thermal and relaxation studies.

EXPERIMENTAL

Molecular weight of the polymers was measured by Waters 510 GPC at 140°C using 1,2,4-trichlorobenzene as a solvent and polystyrene as a standard. The number (M_n) and weight (M_w) average molecular weight, and polydispersity index (PDI, M_w/M_n) are calculated from the GPC curves. Melting and crystallization behaviors of the blends were examined by Perkin-Elmer DSC-7. Indium and zinc were used for the calibration of the melting peak temperature and enthalpy of fusion. The samples were scanned up to 180°C at a heating rate 10°C/min, annealed for 5 min and cooled down to 50°C at a cooling rate 10°C/min. Again re-scanned at the same rate and temperature interval as mentioned before. The melting and crystallization temperature, and heat of fusion and crystallization were calculated from the second scan. The relaxations were measured by using the Polymer Laboratories DMTA Mk III in the range -145°C to T_m -10°C, and tensile mode at a constant frequency of 1 Hz, and at a heating rate of 2°C/min. The mechanical properties were measured using universal testing machine INSTRON 4465 with a cross-head speed 200 mm/min at ambient temperature. All the specimens were dumbbell-shaped with 2 mm thickness, 12.6 mm width and 25 mm gauge length. Izod impact energy was measured by Tinius Olsen impact tester at notched mode at -27°C.

RESULTS AND DISCUSSION

Table 1. Characterization data of the polymers used in this study

	MCPE	HDPE	PP
Melt index, g/10 min	5	5	3.5
Density, g/ml	0.87	0.968	0.9
M_n x 10^4	8.29	2.68	11
M_w x 10^5	1.73	2.6	4.76
PDI	2.09	9.71	4.33

The characterization data of the polymers used are shown in the Table 1. It is well known that polyethylene and polypropylene form compatible blends whereas polyolefins and nylon 6 form incompatible ones though both two types of blends are thermodynamically immiscible. In the case of our polyolefin blend systems, DSC measurement indicates the blends are immiscible due to the existence of two distinct melting peaks. The depression of melting peak is at maximum for HDPE-MCPE system compared to the PP-MCPE which implies that the interaction between polymer segments is maximum for the former system. The melting temperature, crystallization temperature, heat of fusion and heat of crystallization of HDPE and PP are linearly dependent on the amount of the respective polyolefin in the blends. The melting peak shapes containing higher content of MCPE with

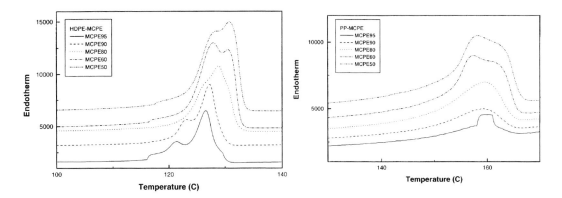

Figure 1. The melting peak shape of HDPE in the second scan of DSC thermograms of HDPE-MCPE blends. The number indicates the percentage of MCPE in the blends.

Figure 2. The melting peak shape of PP in the second scan of DSC thermograms of PP-MCPE blends. The number indicates the percentage of MCPE in the blends.

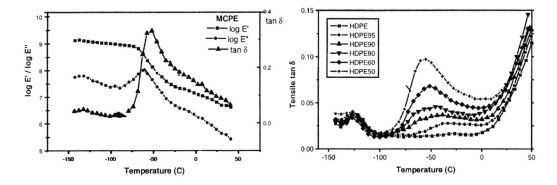

Figure 3. Representative of storage modulus, E', loss modulus, E'', and tensile tan δ as a function of temperature for pure MCPE.

Figure 4. Tensile tan δ spectra of HDPE-MCPE blends and pure HDPE. The number indicates the percentage of HDPE in the blends.

HDPE and PP are shown in Figures 1 and 2 respectively. Although the polydispersity index values are almost comparable but fractionation takes place in the case of HDPE-MCPE blend which also indicates better interaction between HDPE and MCPE.

Figure 3 represents the temperature dependence of tensile storage modulus, E', tensile loss modulus, E'', and tensile tanδ of MCPE which contains 24% 1-octene comonomer. A

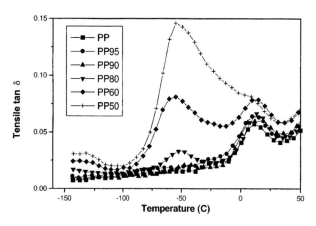

Figure 5. Tensile tan δ spectra of PP-MCPE blends and pure PP. The number indicates the percentage of PP in the blends.

broad relaxation exhibits at about -55°C which is due to the segmental motion of 1-octene comonomer. HDPE does not show any relaxation whereas PP shows at 12°C. Lower content of MCPE (up to about 50%) in PP-MCPE blends show two discernible transitions whereas relaxation shifted to higher temperature as the MCPE content decreases in HDPE-MCPE system as shown in Figures 4 and 5 respectively. So the latter system becomes more compatible than the former ones.

As expected, both the modulus and yield stress decrease as the MCPE content increases in the blends of HDPE-MCPE and PP-MCPE. Tensile strength at break decreases exponentially by the addition of MCPE whereas elongation at break increases linearly up to about 50% of MCPE and then abruptly increases with MCPE for both systems. As expected, impact energy increases with MCPE content in the blend. Impact energy of HDPE-MCPE blend is higher than the PP-MCPE blend.

CONCLUSIONS

Metallocene polyethylene blends with HDPE and PP are immiscible but the degree of compatibility is larger for the former system compared to the latter which reminds the old idea "like dissolves like" is also valid in the polyolefin blends.

ACKNOWLEDGMENTS

S. Choe thanks to the SK Corporation and Inha University for financial support of this work.

REFERENCES

1   A. A. Donatelli, *J. Appl. Polym. Sci.*, 1979, **23**, 3071.
2   S. R. Hu, T. Kyu and R. S. Stein, *J. Polym. Sci., Polym. Phys.*, 1987, **25**, 71.
3   F. M. Mirabella, S. P. Westphal, P. L. Fernando, E. A. Ford and J. G. Williams, *J. Polym. Sci., Polym. Phys.*, 1988, **26**, 1995.
4   J. N. Hay and X. -Q. Zhou, *Polymer*, 1993, **34**, 2282.
5   A. K. Gupta, S. K. Rana and B. L. Deopura, *J. Appl. Polym. Sci.*, 1993, **49**, 477.
6   N. R. Dharmarajan and T. C. Yu, *Plastics Engineering*, 1996, **52**, 33.

7   L. Woo, M. T. K. Ling and S. P. Westphal, *Thermochimica Acta*, 1996, **272**, 171.
8   A. Munoz-Escalona, P. Lafuente, J. F. Vega, M. E. Munoz and A. Santamaria, *Polymer*, 1997, **38**, 589.
9   H. Lee and S. Choe, *Polymer* (Korea), 1994, **18**, 338.
10   K. Cho and S. Choe, *Polymer* (Korea), 1995, **19**, 615.
11   K. Cho, I. Park and S. Choe, *Polymer* (Korea), 1995, **19**, 625.
12   H. Lee, K. Cho, K. Hwang, B. H. Lee and S. Choe, *Polymer* (Korea), 1996, **20**, 317.
13   K. Cho, T. K. Ahn, B. H. Lee and S. Choe, *J. Appl. Polym. Sci.*, 1997, **63**, 1265.
14   H. Lee, K. Cho, T. K. Alm, S. Choe, I. J. Kim, I. Park and B. H. Lee, *J. Polym. Sci., Polym. Phys.*, 1997, **35**, 1633.
15   K. Cho, T. K. Alm, I. Park, B. H. Lee and S. Choe, *J. Ind Eng. Chem.*, 1997, **3**, 147.

Spherulite Boundary Strengthening: A New Concept in Polymer Blends

A. Lustiger, CN. Marzinsky and R. R. Mueller
Exxon Research and Engineering Route 22 East Annandale, NJ 08801

INTRODUCTION

Semi-crystalline polymers, as the name suggests, contain both a liquid-like amorphous phase and an ordered crystalline phase. When solidified from the melt, the crystalline phase resides in platelets called lamellae, which typically grow radially from a central point to form spherulites.

The basis for the phenomenon of spherulitic crystallization is the tendency for the polymer to reject non-crystallizable material during their formation.[1] As a result, the boundaries between adjacent spherulites contain a high concentration of noncrystallizable material composed of low molecular weight chains, atactic polymer, and various impurities. Not unexpectedly, the segregation of non-crystallizable material at spherulite/lamellar boundaries results in mechanical weakness in these regions of the polymer.[2]

The objective of this work is to develop a scheme for increasing interspherulitic and interlamellar strength in polypropylene (PP) while maintaining as much as possible the original crystallinity and melt flow characteristics. The approach we have taken towards this objective is to blend isotactic polypropylene homopolymer with a copolymer that segregates at spherulite boundaries and interlamellar regions like a noncrystallizable impurity during primary crystallization, but then cocrystallizes across adjacent spherulites/lamellae during secondary crystallization. In anthropomorphic terms, the principle is to "fool" the polypropylene homopolymer spherulites into rejecting a copolymer that the homopolymer "thinks" is an impurity. However, after this rejection, the copolymer would strengthen spherulite and lamellar boundaries through forming new connections between them.

EXPERIMENTAL

MATERIALS

Homopolymers

The polypropylene used in this study was an experimental high crystallinity polypropylene (HIX). Through poisoning those catalyst sites which produce atactic PP segments, the crystallinity of IPP can be significantly increased. High crystallinity IPP with an isopentad fraction of 95% was produced with a melt flow rate of 1.5. Molecular weight of the material was M_n= 87,700, M_w=360,900.

Copolymers

Two types of copolymers were used in this study. The first was a 30 MFR ethylene propylene random copolymer (M_w=133,600, M_w/M_n=2.2), containing 5.2% ethylene. The second comprised a multiblock copolymer of isotactic and atactic polypropylene. This copolymer was synthesized based on a paper authored by Chien *et al.*[3] The copolymer had M_w=295,000 and M_w/M_n =2.7.

Blending

Blending of the homopolymers and copolymers took place using a CSI laboratory scale extruder with a barrel temperature of 200°C. In specific cases, 1000 parts per million (PPM) of sodium benzoate was added to these blends in order to raise the crystallization temperature of the homopolymer. Tensile, flexural and impact specimens were compression molded directly from the extrudate at 200°C.

ANALYTICAL CHARACTERIZATION

Fourier Transform Infrared (FTIR) Microscopy

A Nicolet 510P FTIR microscope was used to determine relative ethylene content at spherulite boundaries versus spherulite centers in the homopolymer blended with the random ethylene propylene copolymer.

MECHANICAL CHARACTERIZATION

Flexural testing was accomplished using Instron 4502 mechanical test equipment. Specimens were molded 0.02 inches thick with a gage length of 1 inch. Standard flex and Izod specimens had identical geometries. Flex specimens were tested with a span of 2 inches deformed at 0.05 inches per minute. Flexural data given is an average of three specimens. Izod impact specimens were notched and tested in accordance with ASTM standard D256. Impact data given is an average of five specimens.

RESULTS AND DISCUSSION
RANDOM COPOLYMER BLENDS

To prove the feasibility of the spherulite boundary strengthening concept, it was necessary to experimentally demonstrate that:

1) the homopolymer-copolymer blends would be melt miscible, or at most immiscible on the nanoscale.
2) the copolymer migrates to spherulite boundaries during crystallization of these blends.
3) after this migration, the copolymer forms a link between adjacent spherulites through crystallization of the copolymer chains onto the previously existing homopolymer lamellae in adjacent spherulites.

The first experiment was designed to demonstrate copolymer segregation at the spherulite boundaries. This was done by implementing a classic experiment first described by Keith and Padden.[1] In their experiment, a blend of isotactic and atactic polypropylene was blended in the ratio of 10:90 weight percent and crystallized isothermally using a hot stage. These spherulites were compared with isotactic homopolymer spherulites with no diluent grown under similar conditions.

In their paper, segregation was demonstrated by observing the nature of the spherulites formed in these blends with high diluent (i.e., atactic polymer) concentration. While pure homopolymer yielded the familiar compact spherulites, the high diluent blends displayed open armed, coarse spherulites. These open spherulites are produced because so much noncrystallizable impurity is rejected from the growing spherulites that the formation of new lamellae is sacrificed in favor of thickening of existing lamellae. In contrast, the homopolymer contains a much lower level of impurity, resulting in compact spherulites.

Figure 1 shows a 10:90 blend of PD5052 homopolymer and the random ethylene propylene copolymer described above crystallized isothermally at 137°C, displaying a similar open spherulitic morphology. This appearance confirms segregation of the copolymer to spherulite boundaries.

Cocrystallization

A 75-25 blend of homopolymer and random copolymer was isothermally crystallized at 135°C, etched using potassium permanganate, placed in a constant strain fixture and strained until close to yield. Subsequently, the sample while still in the fixture was placed under the FTIR microscope, and spectra were obtained at the spherulite boundaries and spherulite centers respectively. A composite of these results is shown in Figure 2. The optical microscope results clearly show the presence of oriented material at the spherulite boundaries in the strain

Figure 1. Open spherulitic structure in blend of isotactic polypropylene (iPP) and ethylene propylene (EP) random copolymer.

direction. The orientation suggests that the material that was rejected cocrystallized across adjacent spherulite boundaries, resulting in apparent reinforcement of these boundaries. If cocrystallization had not taken place, the rejected copolymer would not be expected to show such orientation, since it would merely reside at the spherulite boundaries without providing reinforcement. FTIR microscopy reveals a higher concentration of ethylene at the spherulite boundaries than at the spherulite centers as evidenced by the increased intensity of the 730 cm^{-1} peak at the spherulite boundaries. This finding confirms that the ethylene propylene copolymer indeed has a tendency to segregate at these boundaries.

The DSC trace of this blend showed one melting endotherm after isothermal crystallization. However, this blend did not provide cocrystallization under quench conditions due to limitations in the crystallization kinetics, as indicated by the existence of two melting peaks in the DSC. Because practical polymer processing invariably involves highly nonisothermal crystallization, it was necessary to find a copolymer that could cocrystallize with IPP even under quench conditions. In order to develop such a blend, it was presumed that the copolymer must display a much lower crystallization temperature than the IPP. In the case of the IPP-EP blends, the crystallization temperatures were found to be 113°C for the homopolymer and 87°C for the copolymer when cooled at 1°C/minute. This difference in crystallization temperature was clearly not sufficient to provide the necessary cocrystallization under quench conditions.

One obvious way to decrease the crystallization temperature of the copolymer would be to increase ethylene

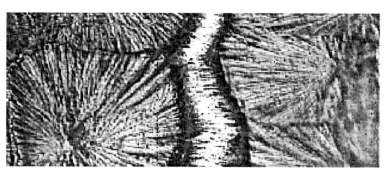

Figure 2. Optical micrograph of strained film of 75-25 blend of iPP homopolymer and EP random copolymer after isothermal crystallization. Direction of strain is horizontal, revealing reinforced spherulite boundary. IR results indicate higher ethylene concentration at spherulite boundaries than at spherulite centers.

content beyond 5.2%. However, based on Lohse's miscibility criteria for EP copolymers,[4] the maximum ethylene content that could be tolerated in a melt miscible blend would be 8%. The change from 5.2% to 8% ethylene would be too insignificant to effect a sufficient difference in crystallization temperature. As a result, a new copolymer was needed.

ISOTACTIC-ATACTIC POLYPROPYLENE (IPPAPP) COPOLYMER BLENDS

As was evident in the Keith-Padden experiment described above, subsequently confirmed by Lohse,[4] atactic polypropylene is melt miscible with isotactic polypropylene in all concentrations. Although pure atactic polypropylene does not crystallize, if one were to use a multiblock copolymer of isotactic and atactic polypropylene (IPPAPP) in a blend with pure IPP, the desired cocrystallization could result.

Such a copolymer has in fact been synthesized by Chien et al.[5,6] using single site catalyst technology. In these papers, polymerization conditions are laid out to synthesize what the authors describe as crystalline-amorphous block polypropylene.

The Keith-Padden experiment was repeated to confirm the appearance of open spherulites in a 90:10 IPPAPP- IPP blend and hence verify segregation of the IPPAPP copolymer to the spherulite boundaries.

Use of Nucleating Agents to Enhance Cocrystallization

A further means of maximizing the difference in crystallization temperatures and hence enhance cocrystallization was to add a nucleating agent to the high crystallinity polypropylene. The nucleating agent chosen was sodium benzoate, incorporated at a concentration of 1000 parts per million (PPM) in the blend.

Modulus and toughness results for the homopolymers and blends are shown below in Table 1. As is well known in the industry, addition of nucleating agent to polypropylene increases flexural modulus but moderately decreases toughness. When IPPAPP alone was added to the HIX PP, toughness increased but modulus decreased. However, when both sodium benzoate and IPPAPP were added, the blend showed a simultaneous increase in both Izod toughness values as flexural modulus. In contrast, if atactic PP (APP) is included as the blending component instead of IPPAPP, a significant reduction in modulus and toughness is evident. Such an effect would be expected with an atactic PP blend, since the desired cocrystallization is precluded.

The simultaneous improvement in toughness and stiffness brought about by the addition of the nucleating agent is presumably because the additive provides both enhanced crystallization which increases stiffness, and enhanced cocrystallization which increases toughness.

In the case of nucleated material, where the spherulites are on the order of one micron in diameter, one would expect that the spherulite boundary reinforcing effect might be an unimportant factor in increasing toughness. However, it should be noted that the cocrystallization

Table 1. Stiffness and toughness of isotactic PP-IPPAPP blends

	Flexural modulus, MPa	Izod impact, J
IPP	1322	0.147±0.006
IPP + Na benzoate	1679	0.133±0.006
IPP + 10% IPPAPP	1250	0.237±0.016
IPP + 10% IPPAPP + Na benzoate	1301	0.289±0.016
IPP + 10% APP	1163	0.128±0.017

effect is presumably not limited to the interspherulitic regions, but takes place between adjacent lamellae as well, resulting in an increase in the interlamellar tie molecule concentration.

CONCLUSIONS

Spherulite boundary and interlamellar strengthening of isotactic polypropylene can take place through blending with an appropriate copolymer. The copolymer migrates to spherulite boundaries as well as interlamellar regions and cocrystallizes, resulting in a higher number of interspherulitic links and tie molecules. Isotacticatactic polypropylene block copolymer has been shown to be an effective spherulite boundary strengthening additive.

Toughness as measured by Izod impact testing has been found to significantly increase through blending with this copolymer. Perhaps more significantly, the toughness enhancement is not accompanied by a significant stiffness decrease as is generally evident in conventional rubber toughening schemes.

REFERENCES

1   H. D. Keith and F. J. Padden, *J. Appl. Phys.*, **35**, 1270 (1964).
2   J. M. Schultz, *Polym. Eng. Sci.* **24**, 770, (1982).
3   D. T. Mallin, M. D. Rausch, Y. G. Lin, S. Dong and J. C. W. Chien, *J. Am. Chem. Soc.*, **112**, 2030 (1990).
4   D. J. Lohse and G. E. Wissler, *J. Mat. Sci.*, **26**, 743 (1991).
5   G. H. Llinas, S. H. Dong, D. T. Mallin, M.D. Rausch, Y. G. Lin, H. Henning-Winter and J. C. W. Chien, *Macromolecules*, **25**, 1242 (1992).
6   B. Rieger, X. Mu, D.T. Mallin, M.D. Rausch and J.CW, Chien, *Macromolecules*, **23**, 3559 (1990).

Modification of Syndiotactic Polystyrene

Modification of Syndiotactic Polystyrene

Modification of Syndiotactic Polystyrene

Jörg Kressler

Martin-Luther-Universität Halle-Wittenberg, Fachbereich Werkstoffwissenschaften, D-06099 Halle (Saale), Germany

Ralf Thomann

Freiburger Materialforschungszentrum und Institut für Makromolekulare Chemie der Albert-Ludwigs-Universität Freiburg, Stefan-Meier-Str. 21, 79104 Freiburg, Germany

INTRODUCTION

The first synthesis of syndiotactic polystyrene (s-PS) was done by Ishihara *et al.* in 1986.[1] s-PS has a complex polymorphism. Three different crystal structures (α, β, γ) and one clathrat structure (δ) have been described.[2-4] Furthermore, the two crystalline modifications α and β are divided in different submodifications having different degrees of structural order ($\alpha', \alpha'', \beta', \beta''$).[2,3] In syndiotactic copolymers of styrene and p-methylstyrene, the α-modification is favored relative to the β-modification.[5,6]

In order to study the influence of longer aliphatic substitution on the phenyl ring on the crystal structure, the crystallization behavior of the copolymers (s-P(S-co-BuS)-X, X $\equiv$ mol% p-n-butylstyrene (BuS)) is studied by means of atomic force microscopy (AFM), wide angle X-ray scattering (WAXS) and differential scanning calorimetry (DSC).[7]

A certain disadvantage of s-PS is the extreme brittleness. There are several possibilities to overcome this problem. It is demonstrated by diffusion experiments employing nuclear reaction analysis (NRA) that s-PP is miscible with deuterated atactic polystyrene (a-PS). The miscibility behavior of s-PS with other polymers is very similar to that of a-PS.

Therefore, a broad range of polymers is available for the modification of s-PS as PPE, Kraton, HIPS etc. The blend morphology of s-PS/Kraton (polystyreneblock-poly(ethene-ran-but-l-ene)-block-polystyrene) is studied by means of atomic force microscopy.

Table 1. Polymer data

Sample	T_g, °C	BuS⁺ content, mol%*	M_n, g/mol**	M_w/M_n
s-P(S-co-BuS)-4	95	4	81000	2.2
s-P(S-co-BuS)-7	93	7	85000	2.2
s-P(S-co-BuS)-11	90	11	124000	1.6
s-P(S-co-BuS)-20	83	20	33000	5.6
s-P(S-co-BuS)-27	82	27	51000	4.1
s-P(S-co-BuS)-40	66	40	18000	6.3
s-P(S-co-BuS)-47	61	47	21000	7.7
s-PS	rrrr-Pantads > 99%		M_n=140000**	M_w/M_n=1.9

* calculated from ¹H- and ¹³C-NMR spectra, ** by size exclusion chromatography versus polystyrene standards

EXPERIMENTAL

MATERIALS

The copolymers of styrene and BuS were polymerized using the metallocene catalyst system $CpTiCl_3$ (CP = cyclopentadienyl)/MAO. Some polymer data are summarized in Table 1. The s-PS homopolymer used was from the BASF AG (Ludwigshafen, Germany). Kraton is a commercially used TPE of Shell.

DSC MEASUREMENTS

Differential scanning calorimetry (DSC) was performed on a Perkin Elmer DSC-7 thermal analyzer.

LIGHT MICROSCOPY

The light microscopic investigations are carried out with an Olympus-Vanox AH2 microscope and a Linkam TMS 90 hot stage that allows observation during isothermal crystallization. The samples used to study the crystalline morphology were prepared by melting the powder of the as-prepared and dried polymer between two cover glasses. The layer thickness between the glasses was about 30 to 50 µm. The samples were held for 10 min at 280°C and then quenched to crystallization temperature with a rate of 99°C/min.

ATOMIC FORCE MICROSCOPY

To study the crystalline morphology (Figure 3) the prepared film was etched to remove amorphous material from the surface. The etching reagent was prepared by stirring 0.02 g potassium permanganate in a mixture of 4 ml sulfuric acid (95-97 %) and 10 g orthophosphoric acid. The 30 to 50 μm thick film was immersed into the fresh etching reagent at room temperature and held there for 1 h. At the beginning, the sample was held in an ultrasonic bath for 15 min. For subsequent washings, a mixture of 2 parts by volume of concentrated sulfuric acid and 7 parts of water was prepared and cooled to near the freezing point with dry ice in iso-propanol. The sample was washed successively with 30 % aqueous hydrogen peroxide (to remove any manganese dioxide present). Then the sample was washed with distilled water. Each washing was supported with an ultrasonic bath. The AFM experiments were carried out with a "Nanoscope III" scanning probe microscope (Digital Instruments) at ambient conditions in the tapping-mode/height-mode (TMHM). The blend morphologies (Figures 5a,5b) were measured in the tapping mode/phase mode (TMPM) without any etching procedure.

WAXS MEASUREMENTS

The measurements were carried out with a Siemens D500 apparatus. For WAXS measurements the CuK_α radiation of a wavelength of $\lambda=0.154$ nm was used.

RESULTS AND DISCUSSION

Figure 1 shows the melting points and T_g values of as-prepared s-P(S-co-BuS)-X samples that are dried under vacuum above T_g to remove any solvent present. The melting points of the copolymers decrease rapidly with increasing BuS content. Melting points range from 270°C for neat s-PS down to 70°C for s-P(S-co-BuS)-40. The neat s-P(BuS) shows a melting point of

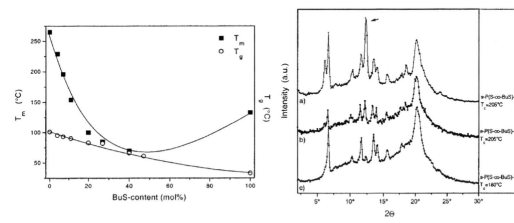

Figure 1. Melting points and glass transition temperatures of the precipitated and dried polymers obtained by DSC as a function of copolymer composition. The lines are drawn to guide the eye.

Figure 2. WAXS traces of s-P(S-co-BuS)-4 isothermally crystallized at 205°C (a) and s-P(S-co-BuS)-7 isothermally crystallized at 205°C (b) and at 180°C (c). The arrow in trace (a) indicates the 040 reflection of the β*-modification.

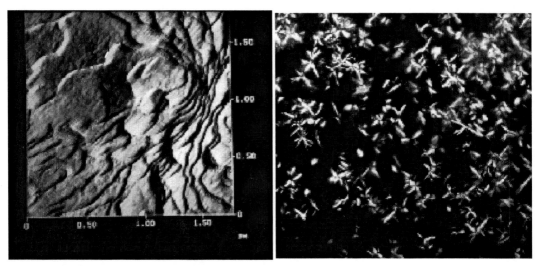

Figure 3. AFM micrographs of s-P(S-co-BuS)-7 isothermally crystallized at 205°C. Crystalline entity formed by broad lamellae (TMHM).

Figure 4. Light micrograph of s-P(S-co-BuS)-7 isothermally crystallized at 205°C (2 cm = 50 µm).

133°C. This indicates the existence of an eutecticum for the melting points as a function of the copolymer composition.[8] Glass transition temperatures of copolymers were found to depend quite linearly on BuS contents. The glass transition temperature for the homopolymer s-P(BuS) is 33°C.

Figure 2 shows the WAXS traces of s-P(S-co-BuS)-4 (a) and of s-P(S-co-BuS)-7 (b) isothermally crystallized from the melt at 205°C. Furthermore, an s-P(S-co-BuS)-7 trace (c) obtained after crystallization at 180°C is shown. Thus the traces (a) and (c) were obtained from samples crystallized at identical supercoolings ($T_m^o - T_c$). s-P(S-co-BuS)-4 isothermally crystallized at 205°C shows mainly the β'-modification. One intense peak of the β'-modification at $2\Theta = 12.3°$ is marked with an arrow. s-P(S-co-BuS)-7 isothermally crystallized at the same temperature shows the α- and the β-modifications. The increasing amount of the α-modification with increasing comonomer content is also observed for syndiotactic copolymers of styrene and p-methylstyrene.[5,6] Crystallized at 180°C, s-P(S-co-BuS)-7 shows mainly the α-modification. This supports again the statement that with increasing content of BuS the α-modification is promoted.

Figure 3 shows an AFM micrograph of an etched sample of s-P(S-co-BuS)-7 isothermally crystallized at 205°C. Crystalline entities formed by lamellae radiating from a central nucleus can be observed. These lamellae have a preferred orientation within the whole supermolecular structure. Therefore, it can be assumed that the supermolecular structure is

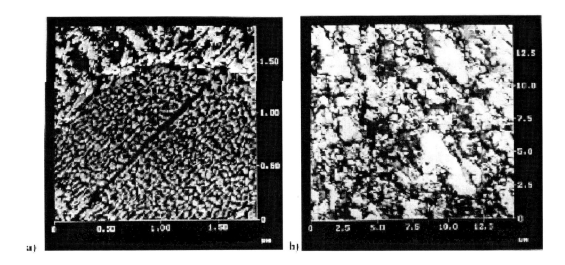

Figure 5. AFM micrographs (TMPM) of s-PS/Kraton blends. (a) s-PS/Kraton 80/20 wt%, (b) s-PS/Kraton 60/40 wt%.

not formed by three dimensional spherulites but from crystalline slices. Different orientations of these slices in the bulk yield the two typical appearances in the light micrograph; similar to spherulites for flat-on slices and elongated entities for edge-on slices seen in the photograph of Figure 4.

Figure 5 depicts AFM micrographs (TWM) of s-PS/Kraton blends. The blends were prepared in a Haake-Büchner kneader at 280°C. Figure 5a shows the AFM micrograph of a 80/20 wt% blend of s-PS and Kraton. s-PS forms the matrix. The block copolymer (Kraton, circular phase) forms well ordered domains. Figure 5b depicts an AFM micrograph of a 60/40 wt% blend of s-PS and Kraton. The minor phase (Kraton) forms the matrix. s-PS particles are irregularly distributed in this matrix. Kraton and s-PS are phase separated. The relatively fine dispersion of the respective phase enclosed in the matrix can be related to the miscibility of the a-PS block of Kraton with sPS. The miscibility of s-PS with d-a-PS is studied by nuclear reaction analysis (NRA). Figure 6 depicts the d-a-PS concentration $\phi(z)$ as a function of depth z, in a two layer sample of s-PS with d-a-PS. The NRA traces were taken after different annealing times at $t_a = 190°C$.

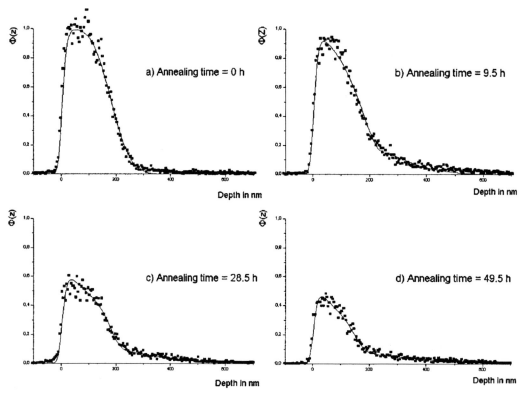

Figure 6. d-a-PS concentration $\phi(z)$ as a function of depth z, in a two layer sample of s-PS with d-a-PS. The NRA traces were taken after different annealing times at ta=190°C.

CONCLUSIONS

Syndiotactic polystyrene is a new engineering plastic. There are several possibilities to over-come the extreme brittleness of s-PS (i.e., the insertion of comonomeric units in the s-PS chain, or the modification of s-PS with tough polymers). Syndiotactic copolymers of styrene with p-n-butylstyrene are used to study the influence of longer substitution on the phenyl ring on the crystallization behavior. Only copolymers with relatively low comonomer content (4 and 7 mol% BuS) are crystallizable from the melt. The copolymers have higher contents of the α-modification than neat sPS at comparable supercoolings and also at identical crystalli-zation temperatures. Crystallized at relatively low supercoolings the copolymers form crys-talline slices built up by broad lamellae.

Blends of s-PS and Kraton are phase separated. The relatively fine dispersion of the respective phase enclosed in the matrix can be related to the miscibility of the a-PS block of Kraton with s-PS. The miscibility of a-PS and s-PS is demonstrated by diffusion experiments employing nuclear reaction analysis.

ACKNOWLEDGMENT

The authors would like to thank Dr Josef Wünsch, BASF, for supplying us with s-PS samples.

REFERENCES

1   Ishihara, N.; Seimiya, T.; Kuramoto, M.; Uoi, M., *Macromolecules*, 1986, **19**, 2465.
2   De Rosa, C.; Rapacciuolo, M.; Guerra, G.; Petraccone, V.; Corradini. P., *Polymer*, 1992, **23**, 1423.
3   De Rosa, C.; Guerra, G.; Petraccone, V.; Corradini, P., *Polym. J.*, 1991, **23**, 1435.
4   Guerra, G.; Vitagliano, M.; De Rosa, C.; Petraccone, V.; Corradini, P., *Macromolecules*, 1990, **23**, 1539.
5   De Rosa, C.; Petraccone, V.; Dal Poggetto, F.; Guerra, G.; Pirozzi, B.; Di Lorenzo, M. L.; Corradini, P., *Macromolecules*, 1995, **28**, 5507.
6   Manfredi, C.; Guerra, G.; De Rosa, C.; Busico, V.; Corradini, P., *Macromolecules*, 1995, **28**, 6508.
7   Thomann R, Sernetz FG, Heinemann J, Steinmann S, Mülhaupt R, Kressler J, *Macromolecules in print*.
8   Wunderlich B, **Macromolecular Physics. Volume 3**, *Academic Press*, New York, London, Toronto, Sydney, San Francisco, 1. Ed. 1980.

Blends of Ethylene/Styrene Interpolymers and Other Polymers: Benefits in Applications

C.F. Diehl, M. J. Guest, B. I. Chaudhary, Y. W. Cheung, W. R. Van Volkenburgh, B.W. Walther

The Dow Chemical Company, Polyethylene and INSITE™ Technology R & D

INTRODUCTION

INSITE[TM] technology has enabled the synthesis of ethylene/styrene interpolymers (ESI) using single site constrained geometry catalysts.[1] The current focus is on ESI containing up to about 50 mole % (~ 80 wt%) copolymer styrene content. These ESI's encompass materials ranging from crystalline to amorphous microstructure, dependent upon the copolymer styrene content.[2,3] Materials engineering through blending of ESI's with other thermoplastic polymers can be used to expand the material performance envelop, and design materials to meet specific performance requirements. A brief background outlining pertinent features of blend technology is given. The characteristics and potential application of selected ESI blends with atactic polystyrene (PS) and polyolefins are then described.

BACKGROUND TO BLEND TECHNOLOGY

MISCIBILITY AND COMPATIBILITY

Depending on the nature and extent of interactions, polymer blends may be classified as miscible, compatible and incompatible.[4] Blends are defined as miscible when the blend components are "molecularly mixed" and form a homogeneous phase. One characteristic is that such a miscible blend exhibits a single glass transition temperature (T_g) intermediate between the T_gs of the pure polymers. It is well known that most polymer blends are immiscible, and have multiple T_gs reflecting the different phases.

An earlier communication described the technology of ESI/ESI blends for ESI's differing in styrene content.[5] The critical comonomer difference in styrene content at which phase separation occurs between two ESI was found to be about 10 wt% for copolymers with mo-

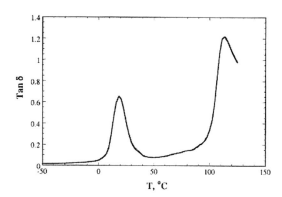

Figure 1. DMS of ES66/PS (70/30) blend (compression molded). Note two distinct T_gs.

lecular weight (MW) of the order of 10^5, both from predictions using copolymer/co-polymer blend models and also from analysis of T_g data for blends and blend components. From such analyses, it is evident that blends of ESI with PS will be immiscible. Figure 1 illustrates the immiscibility of a blend of high styrene content ESI with polystyrene. The blend exhibits two distinct T_gs. Additionally, ESI blends with olefin polymers are expected to be immiscible when the copolymer styrene content of the ESI is above 10 wt%.

Immiscible blends may be further described as compatible or incompatible. Severe phase debonding and significant losses in ultimate mechanical properties are typically found in incompatible blends. Blends showing good mechanical integrity as reflected by the ultimate mechanical properties and diffused interfacial phase boundary are referred to as compatible. Blends of ESI with styrenic polymers, polyolefins and a wide variety of other thermoplastics have been found to exhibit excellent mechanical properties, and typically are compatible.

MORPHOLOGY CONSIDERATIONS FOR THERMOPLASTIC BLENDS

Solid state morphology is a major contributor to immiscible blend properties. Morphologies can include matrix/occlusion types and structures with co-continuity of phases. Blends of interest described herein are all produced in melt processing operations. It is generally accepted (e.g. 6) that factors which determine final solid state morphological structures include the volume fraction of blend components present, interfacial tension between the components in the melt and the shear/elongational flow fields applied to the melt both during blending and also prior to the solidification process for final manufacture of molded parts, films, foams etc.

The underlying approach adopted in our studies has been to pre-design blends dependent upon the desired characteristics of the engineered material, using appropriate morphology/rheology modeling to help the selection of blend components and composition ratios.

EXPERIMENTAL

ESI MATERIALS USED

Composition of ESI is expressed as weight % copolymer styrene. For example, ES70 denotes a copolymer with 70 wt% styrene incorporation. Ethylene-rich and styrene-rich ESI are also referred to as E-series and S-series ESI, respectively. Some of the polymers contain small amounts (< 5 wt%) of PS. The weight average molecular weight (MW) for the polymers generally ranges from about 200,000 to 250,000 g/mol. The MW distributions of all ESI used in this study were narrow (M_w/M_n < 3.5).

BLEND PREPARATION

The melt blends were prepared with a Haake mixer equipped with a Rheomix 3000 bowl. Component polymers were first dry blended and then fed into the mixer equilibrated at 180°C. The molten material was mixed at 180°C and 80 rpm for about 5 min. Test parts produced via compression molding were formed at 190°C.

BLOWN FILM FABRICATION

Films were produced from a dry blend of ES70/PS on a blown film line equipped with a 2.5 inch diameter extruder, a 24:1 single flight, high intensity mixing screw, and a 6 inch diameter high pressure die with 40 mil die gap. Fabrication conditions included a melt temperature of approximately 215°C, a rate of 120 lb/hr, and a blow-up-ratio of approx. 2.5:1. Film thickness was targeted at 2.8 mils.

BLEND SYSTEM EXAMPLES

THERMOPLASTIC BLENDS/MOLDINGS

The broad compatibility of ESI with other common polymers is shown in Tables 1 and 2, which illustrate some of the wide variety of compatible blends that have been produced using ESI. Note that in some cases, E-series ESI demonstrates the best compatibility, while for other polymers S-series is more desirable. For some blend polymers both E-series and S-series ESI can produce a compatible

Table 1. ESI as major component in blends

Major component	Minor component	Property enhancement
E- or S-series ESI	PP, PS, PE	modulus control service temp. range
S-series ESI	PPO	modulus control service temp. range
E- or S-series ESI	SEBS, SBS, SIS	service temp. range elasticity
E- or S-series ESI	EVA, TPU	polarity HF weldability

Table 2. ESI as minor component in blends

Major component	Minor component	Property enhancement
PP	E-series ESI	impact, paintability
PE	E-series ESI	paintability
PS	S-series ESI	toughness, scratch, modulus
PPO	S-series ESI	processability, thoughness
PC	S-series ESI	processability colorability, impact
PVC	E-series ESI/S-series ESI	T_g control, modulus, plasticizing
s-PS	S-series ESI	toughness, ductility
Epoxy vinyl ester	S-series ESI	shrinkage control

blend, depending on the ratio of ESI used, and the relative rheological properties of the blend components.

Table 1 outlines how various performance characteristics of ESI can be enhanced by the addition of a small amount of another thermoplastic polymer. In particular, minor amounts of PS and polyolefins; (particularly polyethylene (PE) and PP) have found wide utility as blend components for modifying modulus and enhancing the heat resistance of ESI.

Figures 2 and 3 demonstrate the blending of 30 wt% of a homopolymer PP or PS into an S-series ESI. These blends were compounded in a Haake mixer and compression molded according to the procedure outlined in the Experimental section. Thermal mechanical analysis was used to measure heat resistance by determining probe penetration depth versus temperature. PP enhances the heat resistance to a greater degree than PS when blended with ESI at the same ratio (Figure 2). An-

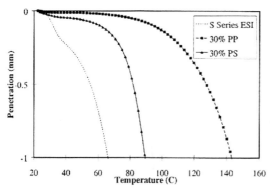

Figure 2. Enhancement of ES70 heat resistance in blends with homopolymers PP and PS, as measured by TMA probe penetration (compression molded samples).

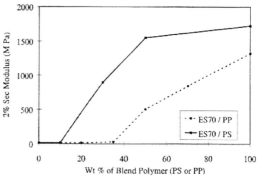

Figure 3. Impact of PS or homopolymer PP level on modulus in binary blends with ES70 (compression molded samples).

other important difference between the use of the selected PS and PP blend components is that PS addition increases modulus, while ESI/PP blends remain relatively flexible. As shown in Figures 2 and 3, the heat resistance of the ES70 can be significantly enhanced by the addition of 30 wt% homopolymer PP, without any significant increase in the modulus, while the addition of 30 wt% PS dramatically increases the modulus of the blend.

PP and PS also differ in their degree of compatibility with the E-series and S-series ESI products. In general, PP is more compatible with the E-series than PS and can be blended at higher levels with the E-series. In contrast, PS exhibits superior compatibility with the S-series than PP, and can therefore be used to produce compatible blends at higher levels of incorporation in S-series ESI.

The specific morphology obtained is highly dependent on the selection of molding conditions in combination with choice of blend components and component ratios. For example, oriented and elongated phase domains have been observed on the surface of injection molded parts prepared from ESI/PS and ESI/polyethylene (PE) blends, and these are considered to affect properties such as modulus and toughness The same injection molded parts typically show a more spherical domain morphology at the core of the part. The concentration of phases remains consistent throughout the part on a macro scale.

FILMS FROM ESI/POLYSTYRENE BINARY BLENDS

ESI has demonstrated potential for film and sheet applications,[7] as a result of such attributes as processability, mechanical properties, heat sealability, stress relaxation, drape, conformability, printability, deadfold properties, elasticity, and thermoformability.

Monolayer and multilayer films have been successfully produced from ESI using a variety of conventional film fabrication equipment including blown, cast, and calendered techniques. Potential applications include labels, window films, decorative overlays, and tapes. Depending on the specific film application requirements, it has often been found advantageous to employ a binary blend of ESI and PS.

The properties of such blends have been found to be strongly dependent on the fabrication process conditions and the resulting blend morphology. For example, a unique morphology has been observed in blown films containing as little as 10 wt% PS, in which very fine (<0.1 μm) elongated PS domains oriented in the machine direction result in films with significantly improved heat resistance, enhanced tensile strength and ambient modulus, while providing a film with good contact clarity, excellent stress relaxation characteristics, and reduced blocking tendencies. Figure 4 compares the dynamic mechanical properties of an ES70/PS blown film to a compression molded sample of the same binary blend.

Figure 5 shows a transmission electron micrograph of a blown film of a ES70/PS. The highly elongated fibrillar PS domains are extremely small (sub-micron in size) and primarily result from the high shear orientation encountered in the blown film process. This unique

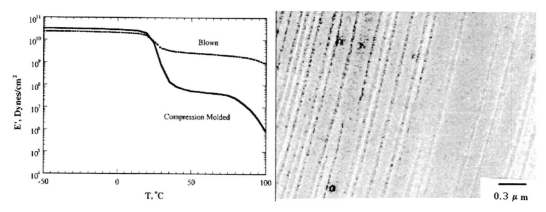

Figure 4. Effect of fabrication induced morphology-properties of an ES70/PS (85/15) blend.

Figure 5. TEM of a ES70/PS (85/15) blend (blown film).

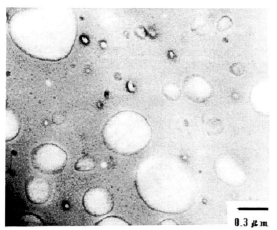

Figure 6. TEM of ES70/PS (85/15) blend (compression molded).

morphology provides high optical transparency. Additionally, the fibrillar morphology provides reinforcement for the blend and significantly increases the modulus of the blend, enhancing its heat resistance, as reflected in the data of Figure 4. In contrast, Figure 6 shows the corresponding compression molded morphology in which the aPS domains appear nearly spherical. Because the domain size is of the order of microns and there is a sufficient refractive index mismatch between ESI and PS, light scattering gives rise to turbidity in the final molded part.

Table 3 compares the blown film properties of a ES70/PS binary blend to those of the neat ES70 and STYRON™ 612. The addition of 30 wt% PS to the ES70 increases the modulus of the film while not significantly affecting the unique tensile stress relaxation or elongation of the ES70. The heat resistance is also positively affected as indicated by the increase in heat seal temperature from 70 to 120°C, while the % haze still remains relatively low.

Table 3. ESI/PS binary blend blown film properties

ESI type	ES70	ES70	ES70	None
ESI, %	100	85	70	0
PS (Styron^{TM} 612), %	0	15	30	100
MD* tensile relaxation, % at 100%	57.4	63.1	63.3	breaks
MD permanent set, % at 100%	10.1	42.1	44.7	breaks
MD 2% secant modulus, MPa	21.4	436.9	782.2	1862
MD yield stress, MPa	4.9	13.1	28.0	40.6
MD ultimate tensile, MPa	40	36	36	41
MD elongation, %	208	210	172	2
Heat seal temperature, °C	70	110	120	N/A
Gardner gloss 45 deg	98	69	71	109
Haze, %	4	8	13	2

* MD= machine direction

FOAMS FROM BINARY BLENDS OF ESI AND POLYOLEFIN OR STYRENIC THERMOPLASTICS

Unique foams have successfully been made by expanding blends of ESI and various thermoplastics, such as PE, PP, PS, and ethylene/vinyl acetate (EVA), using existing polyolefin or polystyrene foam extrusion assets.[8]

Data are presented on extruded foams made from blends of ESI and low density PE (LDPE , 0.923 g/cm^3 density and melt index of 1.8 dg/min (190°C; 2.16 kg). An ES70 polymer with a melt index of 1.5 dg/min was used. Isobutane was used as the blowing agent.

Closed-cell foams with a smooth skin were obtained. Compared with the control LDPE foam, the foams made from LDPE/ESI blends were softer, tougher, and exhibited better drape/conformability. Figure 7 shows that the LDPE/ES70 blend foams also exhibited significantly lower densities than the LDPE control foam, and that the cell sizes of the LDPE/ESI blend foams were significantly smaller than the cell sizes of the reference LDPE foam. Small cells result in a finer cell structure which is aesthetically more desirable.

As shown in Figure 8, the dimensional stability of the LDPE foam was significantly improved by blending the ES70 with the LDPE at a 50/50 ratio. The styrene functionality of the ESI apparently reduces the permeability of the polymer matrix to the isobutane which reduces foam shrinkage.[9] The styrene functionality of the ESI also enhances the printability or paintability of the LDPE/ESI blend foams.

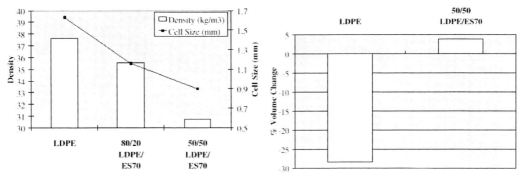

Figure 7. Density and cell size of foams made from LDPE/ES70 blends (7.5 phr isobutane blowing agent).

Figure 8. dimensional stability of LDPE/ES70 blend foams, illustrating the reduction in shrinkage due to addition of ESI.

Modification of PS-based foams with ESI has been shown to result in a softer foam with improved impact toughness and flexibility.[8] Control of open/closed cells is also enhanced in PS/ESI blend foams. ESI's have also been shown to be effective compatibilizers for polyolefins and PS[10,11] and foams have been made from tertiary blends of PE, PS and ESI.

CONCLUSIONS

The performance requirements of potential applications currently under development for ESI can be met by thermoplastic blends with one or more thermoplastic polymers. ESI's exhibit compatibility with a wide range of polymers due to their inherent combination of olefinic and styrenic functionality. Blends of ESI with PS, PE, and PP, in particular, have been found to offer interesting and unique combinations of performance properties. The selection of blend components, blend composition ratio and final part fabrication conditions are key factors in determining structure/property relationships. Appropriate morphology/ rheology modeling is useful in blend design.

REFERENCES

1   **US Patent 5,703,187**, The Dow Chemical Company.
2   Cheung, Y. W., Guest, M. J., Proc. 55th SPE ANTEC, 1634, (1996).
3   Chen, H., et al., *J.Appl.Polym.Sci.*, **70**(1), 109 (1998).
4   Krause, S. J., *Pure Appl. Chem.* **58**(12), 1553 (1986).
5   Cheung, Y. W., Guest, M. J., Proc. 56h SPE ANTEC, 1798 (1998).
6   Utracki, L.A. "Rheology and Processing of Multiphase Systems" in "Current Topics in Polymer Science; Vol. 11", p 30 et seq., Hauser (1987).
7   Murphy, M. W., "Ethylene/Styrene Interpolymer Materials for Film and Sheet Applications", Proc. SP '97, 13'h Ann. World Congress, Zurich, Dec. 1-3, 1997.

8   Karande, S. V., Chaudhary, B. I., FOAMPLAS'98, p. 1-8 in Proc. of 2d Worldwide Conference on Foamed Thermoplastics Markets & Technology, May 19-20, 1998.

9   Briscoe, B. J., Chaudhary B. 1. and Savvas, T., *Cellular Polymers*, **12**, 171 (1993).

10   **US Patent 5,460,818**, The Dow Chemical Company.

11   Park, C. P. And Clingerman, G. P., *Plastics Engineering*, 97-99 (March 1997).

Easy Processing Metallocene Polyethylene

Ching-Tai Lue

Univation Technologies (Exxon Chemical/Union Carbide Joint Venture), P.O. Box 5200
Baytown, TX 77520

INTRODUCTION

The development of metallocene catalysts has been considered as one of the major revolutionary steps in the history of plastic industry.[1] Like the introduction of gas-phase linear polyethylene by Union Carbide in the 1970's, the commercialization of metallocene technology in the 1990's is making a dramatic impact on the polyethylene industry. It is changing the "rules of the game" for polyolefin manufacturing, the polymers produced and their applications.

The first commercial mPE was introduced to the market as "differentiated" or "specialty" products for applications like food packaging and impact modifiers. They are now available with resin densities ranging from 0.86 to 0.91 g/cm^3 and are characterized by narrow molecular weight distribution (MWD) and narrow compositional distribution (CD). These products are valued by the market for their very low extractable, outstanding heat seal performance, high gas permeability, excellent clarity and superior toughness. Typical examples are the Exact^{TM} plastomers by Exxon commercialized in 1991 and the Affinity^{TM} and Engage^{TM} products by Dow in 1993.

In 1995 the first "commodity" mPE products were commercialized by Exxon as Exceed^{TM} polyethylene. Unlike its predecessor Exact^{TM} plastomers which are made initially by a high pressure process and recently by solution and slurry processes as well, the Exceed^{TM} products are made by the more cost effective UNIPOL® gas-phase process. With densities ranging from 0.915 to 0.930 g/cm^3, they were introduced to the market to compete with ZN-LLDPE's for applications such as can liners and stretch film. These products maintain the narrow MWD/CD features and provide advantages in toughness and optical properties.

Between 1995 and 1997, numerous resin suppliers publicly disclosed their development efforts and intentions to commercialize mPE.[2,3] These companies, in alphabetical order, include BASF, Borealis, BP, Dow, DSM, Fina, Mitsui, Mobil, Phillips, Quantum (now Equistar

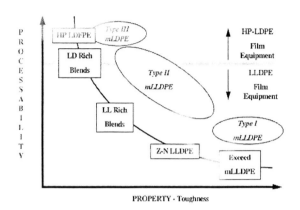

Figure 1. New generation of metallocene PE.

Chemicals) and Union Carbide. Commercializations in this period include Elite[TM] (from Dow), Evolue[TM] (from Mitsui), Borecene[TM] (from Borealis) and Luflexen[TM] (from BASF) polyethylenes. All claimed to have enhanced performance over conventional PE products.

Currently, a wide array of mPE products is being developed in the industry to possibly replace the whole spectrum of non-functionalized conventional PE products.[4,5,6] These products are grouped in three families as shown in Figure 1.

Unlike the early generation of mPE, where narrow MWD and narrow CD are the keys for its outstanding performance, the up-coining new mPE products are focusing on the "control/manipulation" of MWD, CD, long chain branch (LCB) and long chain branch distribution (LCBD). Manipulating these molecular structures to meet a specific set of performance requirements has become a popular theme for future mPE development.

In this paper, a new family of easy processing metallocene polyethylene (EZP-mPE) will be presented. The resins exhibit a unique combination of easy processing, outstanding toughness and good optical properties that has not been seen with existing PE products. This new family of mPE has a great potential for applications where blends of ZN-LLDPE and HP-LDPE are currently used.

EXPERIMENTAL

The EZP-mPE in this study was an experimental ethylene/hexene (E/H) copolymer with target MI and density of 1.0 dg/min and 0.918 g/cm^3, respectively. For comparison, a series of blends was prepared from a selected pair of commercial ZN-LLDPE and BP-LDPE at various ratios. These blends were prepared with a WP-57 twin screw extruder. The ZN-LLDPE was also an E/H copolymer. The BP-LDPE was a homopolymer. Table 1 lists the materials and test results.

Blown film samples were made on a 2 1/2" (64 mm) Egan line with a typical LLDPE Sterlex barrier screw, a 6" (150 mm) diameter die and a Uniflo dual lip air ring. The die gap was set at 30 mil (0.8 mm); target film gauge was 2 mil (50 μm); blow up ratio was 2.5 and output was about 115 lb/hr (52 kg/hr).

Table 1. Summary

Product	100% ZN-LL	75% ZN-LL 25% HP-LD	50% ZN-LL 50% HP-LD	25% ZN-LL 75% HP-LD	100% HP-LD	EZP-mPE
Density, g/cm3	0.919	0.920	0.921	0.921	0.921	0.919
MI, dg/min	1.0	0.8	0.7	0.6	0.6	1.1
FRR (I-21/I-2)	29	33	41	53	76	46
64 mm Egan blown film line (0.8 mm die gap, 2.5 BUR, 50 micron)						
Output, kg/h	51	52	49	51	52	52
Head press., MPa	21.3	20.2	18.2	16.3	13.3	15.6
Motor load, amp	39.6	38.7	33.8	30.3	24.6	30.9
MD tensile, MPa	46.7	38.1	29.8	28.4	28.0	44.1
TD tensile, MPa	47.1	37.7	33.2	27.5	25.1	45.0
MD tear, g/micron	17.8	4.2	2.8	3.4	5.4	8.2
TD tear, g/micron	26.4	26.0	20.5	8.4	3.3	16.9
Dart impact, g/micron	10.2	4.6	4.1	3.7	3.8	16.1
Haze, %	14.7	8.8	8.6	8.1	6.7	10.2

Film properties were measured according to ASTM methods after at least 40 hours of conditioning. The dart impact strength reported in Table 1 was measured from a 26" (66 cm) height (Method-A).

DISCUSSION

As suggested by FRR (flow rate ratio at condition-F vs. condition-E according to ASTM D1238), the EZP-mPE was inherently more shear thinning than the conventional ZN-LLDPE (46 vs. 29, respectively, in Table 1). One would therefore expect this new EZP-mPE to process more easily than a conventional ZN-LLDPE. Using extrusion pressure (Figure 2) and motor load (Figure 3) as measurements for processability, the EZP-mPE is about 20-30% better than the ZN-LLDPE at a similar MI. This is very different from the currently available commercial mPE's whose processability is often inferior to or, at best, comparable to that of a ZN-LLDPE product. This new family of EZP-mPE is truly an easy processing product. Compared with LL/LD blends, it is as easy to process as a LD-rich (25% LL/75% LD) blend (see Figures 2 & 3).

For tensile strength, EZP-mPE is comparable to the ZNLLDPE and significantly better than all of the blends (Figure 4). The most impressive single feature of this product is its exceptionally high impact strength: 60% higher than the ZN-LLDPE and about four times that of ZN-LL/BP-LD blends (see Figure 5). As for the Elmendorf tear, EZP-mPE is significantly

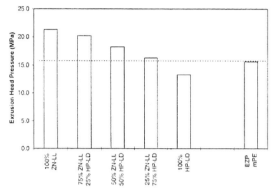

Figure 2. Comparison of extrusion pressure.

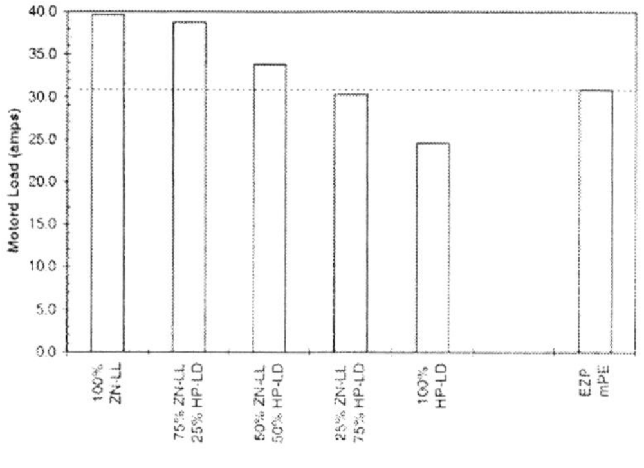

Figure 3. Comparison of motor load.

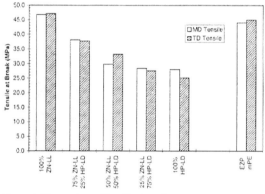

Figure 4. Comparison of tensile strength.

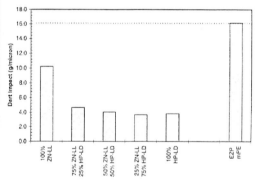

Figure 5. Comparison of dart impact.

better than all of the blends in both MD tear and MD/TD balance (Fig-6). The balance in MD/TD tear is important for a film to avoid "splitiness", i.e., a great tendency to fail in one particular direction.

In the history of PE product evolution, there has always been a trade-off between film properties and processability. From the above discussion, this paradigm has been broken by this new family of EZP-mPE. It presents a tremendous potential to replace a wide range of conventional PE products and their blends.

Figure 7 and Figure 8 are radar charts that illustrate this possibility. The charts were constructed to show the differences between products. The properties have been grouped as

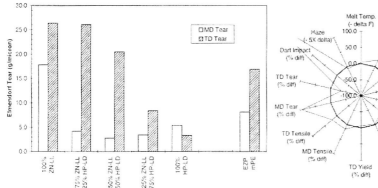

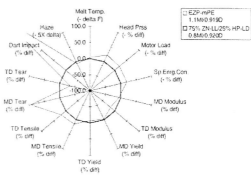

Figure 6. Comparison of Elmendorf tear.

Figure 7. Replacement for LL-rich blends.

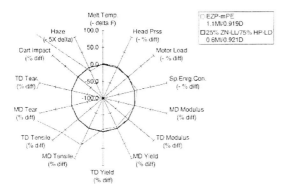

Figure 8. Replacement for LD-rich blends.

follows: In the clockwise direction, processability (4 arms), stiffness (4 arms), toughness (5 arms) and optical property (1 arm). There is a reference LL/LD blend on each chart whose performance is normalized to a value of "0". The scale on each arm shows the "relative difference" between the EZP-mPE and the reference LL/LD blend. A point moving away from "0" means "better" or "improving" over the reference blend.

Figure 7 illustrates the opportunity for the EZP-mPE to replace a LL-rich blend (75%LL/25%LD). It shows that the EZP-mPE can be processed more easily than the reference blend and, at the same time, provide a film which is much tougher. There is essentially no difference in stiffness between the two products. The advantages that EZP-mPE has to offer over LL-rich blend are clear. The converters will benefit not only from the easier processing but also from the downgauging opportunity for cost savings. The end users will also benefit from a stronger film for performance and from downgauging for environmental concerns.

When comparing to the LD-rich (25%LL/75%LD) blend, the advantages of EZP-mPE in toughness are even more dramatic (Figure 8). On an extrusion line such as the one used in this study, there would be no observable loss in processability.

CONCLUSIONS

The new family of EZP-mPE has shown a unique combination of product attributes not seen before. It changes the "rules of the game" for the balance/combination among processability, stiffness and toughness of PE products. It offers an excellent opportunity in applications where blends of ZN-LLDPE and BP-LDPE are currently used. It is as strong as a conventional ZN-LLDPE for most of the toughness measures and as easy to process as most of the LL/LD blends. There is a great potential for downgauging which will benefit not only the converters, but also the consumers with environmental concerns and overall cost reduction needs.

ACKNOWLEDGMENTS

The author recognizes contributions from N. A. Merrill and G. A. Vaughan for catalyst development, M. E. Muhle for making the experimental mPE product and S. Ohlsson for valuable inputs.

REFERENCES

1   V. Wigotsky, "Metallocene: The Next Step", *Plastics Engineering*, P. 16 (July 1995).
2   MetCon'95; Worldwide Metallocene Conference 1995.
3   Plastics Formulation and Compounding; (Dec. 1995).
4   W. A. Fraser, J. L. Adams & D. M. Simpson, "Polyethylene Product Capabilities from Metallocene Catalysts with the UNIPOL® Process", 4h International Conference on Metallocenes Polymers: Metallocenes Asia '97, May 1997.
5   D. M. Simpson, P. D. Whaley & H. T. Liu, "Product Performance of mLLDPE Resins with Improved Processability", SPO Conference in Houston (1997).
6   G. N. Foster, J. L. Adams, D. C. Lee, R. H. Vogel & S. H. Wasserman", "Advanced Polyethylene Product Design for Technology Licensing", Advances in Polyolefins, Sept. 1997, Napa, California, Sponsored by ACS.

Effect of Metallocene Polyethylene on Heat Sealing Properties of Low Density Polyethylene Blends

Juan D. Sierra

Instituto de Capacitación e Investigación del Plástico y del Caucho, ICIPC Medellin, Colombia, P.O. Box 2629

María del Pilar Noriega and Tim A. Osswald

Polymer Processing Research Group, University of Wisconsin - Madison, USA.

INTRODUCTION

The heat sealing is the typical technique used in modern machinery for packaging of liquid and solid products. Due to the good cost/performance balance of low density polyethylene (LDPE) this polymer is widely used as a sealing layer in several laminations for packaging of diverse products. The increasing demand for higher speed in VFFS and HFFS packaging equipment and better integrity, has imposed surplus requirements for polymers, including better hot tack values, low initiation temperature of sealing and wider temperatures ranges of sealing. The mentioned features are quite difficult to meet with traditional polymers so it is mandatory to consider alternative materials with better sealing properties and higher prices.

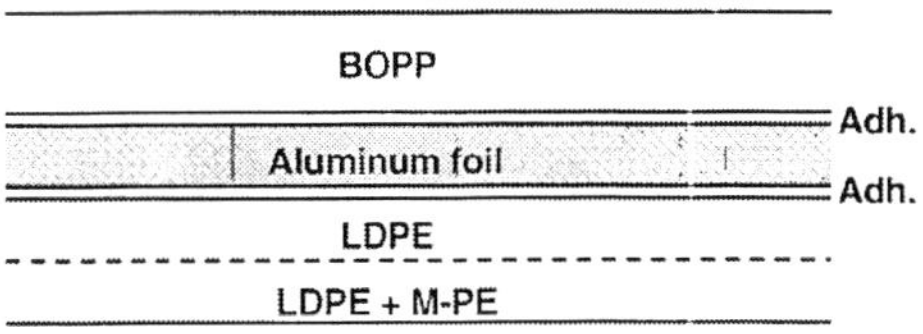

Blending of traditional LDPE with polymers with better sealing properties, such us m-PE and EVA,[1,2] is a good alternative to meet the higher speed demands in modern packaging machinery. Even blends with low quantities of m-PE are enough to modify significantly the heat sealing properties of LDPE. The purpose of this research is to evaluate the percentage of m-PE (based in octene comonomer) for optimal balance of sealing properties and cost in laminations of BOPP/aluminum foil/PE sealing layer for packaging of diverse products (see Figure 1).

Figure 1. Details of laminate structure.

The interdiffusion theory suggests that during the sealing process of semicrystalline polymers the following phenomena occur sequentially:[3,4]

- **Melting of the polymers**: low melting temperature and low heat of fusion enhances sealability.
- **Wetting of the interfaces**: reduction of the distance to maximize intermolecular attraction forces. A proper compatibility at the interface is very important for good seal integrity.
- **Diffusion of chains segments across the interface**: for best seal strength a great degree interdiffusion of chain segments is required.
- **Recrystallization of some of the chain segments**.

According to the mentioned interdiffusion theory, for a good seal the following properties are required:

- Enough heat transfer that guarantees proper melting and mobility of chain segments at the interface.
- Intimate contact of both surfaces.
- Enough time and compatibility for a high degree of interdiffusion of chain segments across the interface.
- Not too high crystallization degree which makes a fragile seal.

EXPERIMENTAL

The tested materials were a high-pressure low density polyethylene (LDPE) of MFI of 2,0 for film extrusion and a metallocene polyethylene (m-PE) with 12% octene comonomer and MFI 1,0 as a sealing modifier (see Table 1).

The present study is based on measurements of heat seal strength and hot tack values. Complementary studies of glass transition and melting point were done using a Differential Scanning Calorimeter, DSC. The following instruments were used:

Table 1. Information on evaluated polymers

	High-pressure, low-density polyethylene, LDPE	Metallocene polyethylene, m-PE
Producer	Ecopetrol, Colombia	Dow Plastics
Polymerization process	High-pressure	Metallocene catalyst
Comonomer type		octene
Comonomer content		12%
MFI, g/10 min at 190°C and 2.16 kg	2.0	1.0
Additives	slip: Erucamide 700 ppm, antiblock: 1260 ppm	

- HSG/ETK Brugger Feinmechanik for hot tack (falling weight type) data and sample preparation for seal strength measurements.
- Instron model 4464 universal testing machine with pneumatic clamps and a precision 100 N load cell.
- Blends of m-PE and LDPE were prepared using a Torque Rheometer with following specifications: Haake model Rheocord 90 with Rheomix 600 mixing chamber and roller rotors.
- Differential Scanning Calorimeter, DSC, TA Instruments, 2910, a heat flow type calorimeter, with a standard cell.

Several heat seal strength and hot tack measurements were done on laminations of BOPP/aluminum foil/PE sealing layer with amounts between 10 and 33% weight of m-PE prepared using an industrial lamination machine. Every reported point is the average of at least ten individual measurements.

For the DSC analysis, blends of m-PE and LDPE similar to those used in lamination process were prepared using a Torque Rheometer at following conditions: $180°C$, 60 rpm, 4 min and filling factor of 70%. Measurements of the glass transition temperature of the raw materials and the blends were done. A rapid quenching to enhance amorphous phase was performed on samples around 10 mg which were heated at $5°C/min$ from -100 to $0°C$. Additional DSC melting point measurements were done at $5°C/mm$ heating rate. A preliminary run was done to erase all thermal history.

RESULTS AND DISCUSSION

HEAT SEAL STRENGTH

The seal strength as a function of the programmed temperature of seal clamp for blends containing 10, 15, 20 and 33% weight of metallocene polyethylene (m-PE) are presented in Figure 2.

As the content of metallocene polyethylene (m-PE) increases from 10 to 33% the temperature to obtain the maximal values of seal strength decreases from 125 to $105°C$. The temperature to obtain the maximal values of seal strength for the 15 and 20% blends was around $110°C$. An optimal value for seal strength of 21.0 N/15 mm was obtained for 15% blend. The seal strength for the 20% blend was very close to the optimal.

HOT TACK VALUES

The hot tack value as a function of the programmed temperature of seal clamp for blends containing 10, 15, 20 and 33% weight of metallocene polyethylene (m-PE) are presented in Figure 3. The hot tack is reported as the length of seal that remains intact under a certain weight hanged immediately the seal is done.

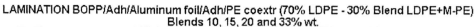

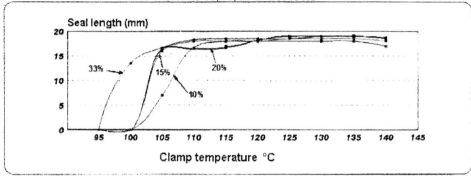

Figure 2. Heat seal strength for blends of m-PE and LDPE.

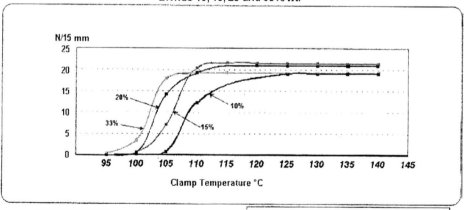

Figure 3. Hot tack values for blends of m-PE and LDPE.

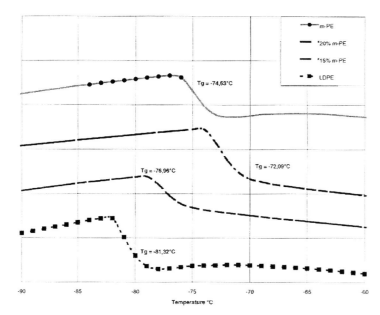

Figure 4. Glass transition of m-PE/LDPE and blends.

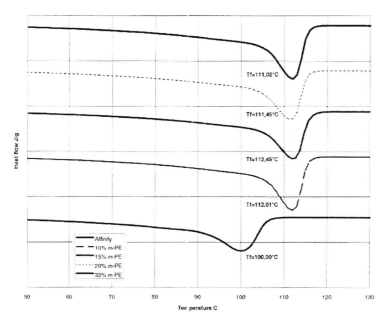

Figure 5. Melting point of m-PE/LDPE and blends.

The content of metallocene polyethylene (m-PE) has a dramatic effect on the temperature to obtain acceptable hot tack values. The blend of 33% metallocene polyethylene (m-PE) exhibited high hot tack at lower temperatures but the maximal value was observed at higher temperatures. Blends of 15 and 20% exhibited a similar behavior, characterized by a rapid increase of hot tack around 100°C and full development of the hot tack around 115°C.

DIFFERENTIAL SCANNING CALORIMETERY

A unique value of glass transition of blends between the values observed for m-PE and LDPE allows us to conclude that both resins have total compatibility for the compositions studied (see Figure 4). The total compatibility of both resins guarantees not only the possibility of high degree of interdiffusion of chain segments across the interface (excellent seal properties) but also a good appearance of the lamination.

A melting point between the values observed for m-PE and LDPE was obtained for the blends, as well as intermediate heat of fusion (see Figure 5). These observations confirm the high compatibility and allow us to anticipate better heat sealing performance of these blends.

CONCLUSIONS

Blending of traditional high-pressure low density polyethylene (LDPE) with metallocene polyethylene (mPE) enhances the heat seal properties of sealing layers in laminations of BOPP/aluminum foil/PE sealing layer. Even low content of metallocene polyethylene (m-PE) significantly improves the sealing properties of the PE sealing layer by allowing to obtain a good seal strength and hot tack values at lower temperatures. As a benefit of the enhanced heat sealing performance this lamination could meet the modern demands for higher speed in VFFS and HFFS packaging equipment and a better integrity of package.

As a result of an extensive study with blends of high pressure low density polyethylene (MFI=2.0) with metallocene polyethylene (12% octene and MFI=1,0) as a sealing layer in laminations with BOPP and aluminum foil, it was found that a blend around 15% has the best balance of seal properties and cost.

DSC studies of glass transition and melting point of the blends allow us to anticipate total compatibility of both resins and good sealing properties based on the possibility to study interdiffusion of chain segments across the interface.

NOMENCLATURE

BOPP	Bioriented Polypropylene
DSC	Differential Scanning Calorimeter
EVA	Ethylene Vinyl Acetate Copolymer
HFFS	Horizontal Form Fill Seal machine
LDPE	Low Density Polyethylene
m-PE	metallocene polyethylene
MFI	Melt Flow Index
PE	Polyethylene
VFFS	Vertical Form Fill Seal machine

ACKNOWLEDGMENTS

The authors gratefully acknowledge the support of this research by Microplast S.A. and by the research teamwork of the ICIPC: Carmiña Gartner, Silvio Ospina and Juan Carlos Posada.

REFERENCES

1   R. M. Patel and others. Comparison of EVA and Polyolefin plastomer as a blend component in various films applications. ANTEC 97.

2   H. Shih and others. Effect of Polyethylene blend on heat sealing properties. ANTEC 98.

3   Mueller C., Hiltner A., Baer E. Interdiffusion and crystallization of LLDPE: relationships to heat sealability. ANTEC 96.

4   Nicastrol C. and others. Change in crystallinity during heat sealing of cast polypropylene film. *J. Plastic Film & Sheeting*, **9**, No. 2, April 1993. P. 159-167.

Extrusion Operation Window for Filled Metallocene Polyethylenes

María del Pilar Noriega and Tim A. Osswald

Polymer Processing Research Group Department of Mechanical Engineering University of Wisconsin-Madison Madison, Wisconsin 53706

Omar A. Estrada

Instituto de Capacitación e Investigación del Plástico y del Caucho (ICIPC) Medellín, Colombia

INTRODUCTION

Several authors have conducted independent experimental work with barrier screws for linear low density polyethylene and metallocene resins for a single screw extruder developing conclusions which are specific to their choice of screw design, processing conditions, materials and instrumentation.[1-4] The intention of this research is to compare at first the mPE material with respect to LDPE based on the characteristic curves and energy balance, afterwards, to compare different blends of mPE/MB with respect to a reference blend (mPE/LLDPE) without the filler (TiO_2). Results were obtained under the same processing conditions and screw design. The comparison involves the analysis of the characteristic curves, pressure production or consumption along the screw, temperature homogenization and mechanical properties for each blend.

EXPERIMENTAL

The experiments were performed on a highly instrumented Extrudex ED-N-45-25D single screw extruder. The extruder was equipped with a 45 mm barrier screw (9D- Feed zone/11D-Barrier/5D- Metering) with a 2.5:1 compression ratio. No mixing device was fitted to the screw to enable us to assess the viscous dissipation behavior of the metallocene material.

Before experimentation, the materials were subjected to rheological measurements performed at the ICIPC. The material and data are as follows,

1. Metallocene polyethylene (mPE - Affinity 1881) with Carreau indices of
    A = 3115 Pa.s, B = 0.0835 s and C = 0.47 at 237.5°C.
2. Linear low density polyethylene (LLDPE - Ipako 6200-1) with Carreau indices
    of A = 2580 Pa.s, B = 0.0407 s and C = 0.52 at 237.5°C.
3. Masterbatch (MB) of titanium dioxide at 50 % in weight with linear low density
    polyethylene as the carrier.

The trials were conducted under the same heating band temperature profile (270, 260, 250, 240, 230°C) with four different die openings and three screw speeds for a total of twelve operating points per trial. The performance of each blend was evaluated by means of the characteristic curve and pressure production or consumption. The pressure was measured with five pressure transducers located at 12D, 16D, 20D, 22D and 25D along the extruder barrel.

The extrudate was a mix of a titanium dioxide master batch (MB) with the respective polymer. The melt temperature profile measures the thermal homogeneity. The profile was measured with the use of a five sensor thermal comb located 2D after the end of the screw.

The mechanical properties of the blends were evaluated based on injection molded specimens.

RESULTS

SCREW CHARACTERISTIC CURVES

Depending on the type of blend utilized, the operating window of the extruder will differ. The extruder performance is not only controlled by the polymer material, but is dependent on the screw design from the hopper to the die. The screw characteristic curves shown in Figures 1, 2 and 3 reveal the extruder's operation range.

Figure 1 shows the wider operation window of the metallocene resin compared to the conventional low density polyethylene. The mPE processing involves higher output, higher back pressure and torque than for the conventional LDPE.

For the operating range tested, the blend mPE + 5 % MB gives, in general, the lowest output characteristics. Note how the masterbatch content of 5 % increases slightly the output compared to the reference blend (mPE + LLDPE) and compared to mPE 100 %. Figure 2 shows the operation window of the blend and its reference.

The blend mPE + 15 % MB gives, in general, the highest output characteristics. The masterbatch content of 15 % increases remarkable the output and the backpressure compared to the reference blend (mPE + LLDPE). The curve is completely shifted up to the high output region. Figure 3 shows the operation window of the blend and its reference.

When comparing the operating points with mPE + MB (Figs. 2 and 3) and mPE + LLDPE it is seen that the mass throughput for mPE + MB is always greater than the reference blend. This fact is related to the different bulk density and rheology of the two materials. The

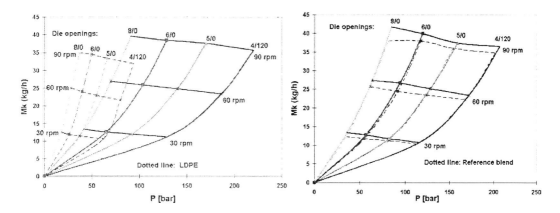

Figure 1. Screw characteristic curves of LDPE vs. mPE.

Figure 2. Screw characteristic curves of mPE + 5% MB vs. reference blend (mPE+LLDPE).

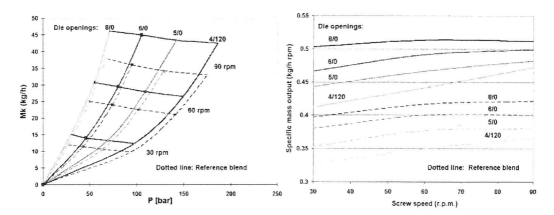

Figure 3. Screw characterisitc curves of mPE + 15% MB vs. reference blend (mPE+LLDPE).

Figure 4. Specific mass output of mPE + 15% MB vs. reference blend.

MB used in the experiments has a higher bulk density and a lower viscosity than the mPE at the operating conditions.

The specific mass output [mk (kg/h)/N (rpm)] of the extruder increases at higher screw rotation speeds and at higher filler content as expected (Figure 4).

Table 1. Energy balance LDPE vs. mPE for all the operation points

W*	N	LDPE				mPE			
	rpm	e_m, J/s	ΔH, J/s	P_p, J/s	%E, J/s	e_m, J/s	ΔH, J/s	P_p, J/s	%E, J/s
8/0	30	408.4	2028.4	8.3	499	754.0	2544.3	20.6	340
8/0	60	1193.8	4014.4	25.6	338	2324.8	5134.8	68.8	224
8/0	90	2261.9	5526.3	47.2	246	4241.1	7610.7	131.7	183
6/0	30	377.0	1945.2	113.	519	754.0	2412.5	29.9	324
6/0	60	1131.0	3904.1	35.1	348	2324.8	4992.8	98.5	219
6/0	90	2167.7	5526.3	63.0	258	4335.4	7463.7	181.1	176
5/0	30	377.0	1867.9	15.4	500	691.2	2301.1	40.4	339
5/0	60	1131.0	3747.5	46.2	335	2324.8	4795.1	126.9	212
5/0	90	2167.7	5358.9	80.6	251	4335.4	7346.3	233.6	175
4/120	30	408.4	1729.9	24.0	429	754.0	2147.7	52.8	292
4/120	60	1131.0	3513.7	63.2	316	2324.8	4551.6	159.1	203
4/120	90	2167.7	5155.2	109.9	243	4335.4	7026.9	287.1	169

*W=die opening

ENERGY BALANCE

The energy balance of the extruder is a very important parameter to be considered in the processing of high viscous polymers. The energy efficiency is defined as the ratio between the energy absorbed by the polymer (enthalpy change + pumping capacity) and the energy supplied by the motor (= adiabatic assumption).

$$\% E = [(\Delta H + Pp)/em]100$$

The Table 1 shows the energy efficiency for LDPE vs. mPE for all the operation points. It is seen from the table that the actual barrier screw is efficient at the highest screw speed and die restriction. For all the operation points, the polymer was absorbing energy from the motor and from the heaters (%E >> 100 %).

PRESSURE PROFILE

The pressure profile along the screw in Figure 5 shows how much pressure is required to overcome the die restriction. The figure shows the pressure at the greatest speed, 90 rpm, at the most restrictive die and at the highest filler content. This point was chosen because it is where the screw configuration operates more efficiently. The pressures for the blend mPE +

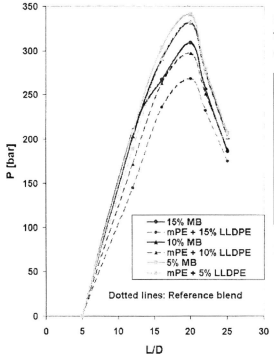

Figure 5. Extruder pressure profile at highest output (90 rpm, min. die opening) for blends mPE + 5%, 10%, and 15% MB.

Table 2. Temperature results at 90 rpm averaging the 4 different restrictions and pressure at the highest output

Blend	T_b °C	σ %	ΔT_{max} °C	ΔP bar
mPE	258	9	197	5.38
mPE+5% MB	256	7.25	167	4.79
mPE+5% LLDPE	256	6.75	153	4.70
mPE+10% MB	255	6.75	153	4.72
mPE+10% LLDPE	255	5.75	134	4.49
mPE+15% MB	253	3.75	146	4.31
mPE+15% LLDPE	253	4.5	109	4.28

LLDPE had MI=20

15 % MB are higher than those for the reference blend due to the different viscosities of the materials under these conditions. Same trend was observed for the blends with 5 % MB and 10 % MB.

For all the trials, the maximum pressure occurs at about L/D = 20 at the end of the barrier section and beginning of the metering section. This maximum pressure values decrease at higher filler content for a constant screw rotation speed due to the different rheology of the blends.

Table 2 shows the pressure loss from 20D till 25D, which correspond to the metering section. The mPE 100% demands the highest pressure because of the highest viscosity. The blend mPE + 15 % MB exhibits the lowest pressure consumption in Table 2. The pressure loss with the different blends is reduced further as the die restriction is increased, thus the extruder becomes more efficient.

MELT TEMPERATURE PROFILE

Figure 6 shows the average melt temperature profile measured by the thermal comb at 90 rpm, at the most restrictive die and at the highest filler content. The temperature at the center is higher for all the polymer blends due to the viscous dissipation effect, which is quantified in

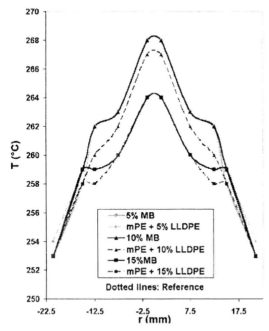

Figure 6. Average melt temperature profile at 90 rpm and min. die opening for blends mPE + 5, 10, and 15% MB.

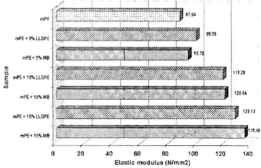

Figure 7. Elastic modulus of mPE + 5, 10, 15% MB vs. reference blends.

Table 2 by the bulk temperature. Due to the smaller gap in the metering section, the material is subjected to a higher degree of viscous dissipation, which increases with the screw speed. The maximum temperature difference in Table 2 between the wall temperature and the center is 7.25°C for mPE + 5 % MB and 3.75°C for mPE + 15 % MB.

The melt temperature homogeneity can be quantified by looking at the maximum temperature difference and the standard deviation of the temperature measurements in Table 2. The blend mPE + 15 % MB results in the most homogeneous temperature profile and lowest viscous dissipation. As expected, the melt homogeneity results show the need of a screw mixing head for the processing of metallocene resins.

MECHANICAL PROPERTIES

Figure 7 presents a bar chart comparing the elastic modulus (ASTM 638M) of the different blends with respect to the reference blend. The highest modulus is obtained for the blend mPE + 15 % MB due to the highest filler content as well. It is seen that the filler (TiO_2) has a reinforcing effect on the blend. This is in agreement with the literature on effect of fillers and reinforcements.[5] The lowest elastic modulus is obtained for mPE. 100%.

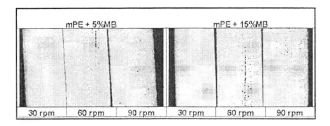

Figure 8. Extrudate surface micrographs for the blends mPE + 5% MB and mPE + 15% MB.

EXTRUDATE MICROGRAPHS

The final product quality can be appreciated from the micrographs in Figure 8. At the highest screw speed (90 rpm) and lowest filler content, the extrudate exhibited visible outer surface irregularities. Fortunately, the interior surface, due to somewhat reduced shear rates, showed a smooth appearance. This is in agreement with work done on processing of metallocene polyolefins by Khare *et al.*[6]

DISCUSSION

The obtained operation windows for the metallocene resin and the blends are excellent to fit the demands of the industry. The mPE processing involves higher productivity and energy requirements than a conventional LDPE as shown in this paper.

It is important to keep in mind that the viscous dissipation problem will get worse as scale-up is involved. Thus, when selecting a metallocene resin and a screw design for a specific application, the problem of viscous dissipation should always be considered. However, it is observed that the melt temperature is more uniform at high level of filler content and die restriction. This effect is related to the rheology of the blend and the average residence time in the extruder.

The barrier screw characterized as low- shear and high efficiency design is the most suitable for the processing of high viscous metallocene materials as recommended in the literature,[2] but a mixing head is required to obtain an adequate melt homogeneity.[7] To satisfy the mixing requirements by the metallocene processing, it could be necessary to use two types of mixing devices in series in single screw extruders, one dispersive and one distributive mixer at the screw tip.

The filler masterbatch in the blend has contributed to a decrease in viscosity due to the LLDPE carrier and positive reinforcement effect. This is desirable for many applications in industries such as film and sheet extrusion.

There is still a point to improve in the metallocene processing related to the surface appearance of the extrudate. These outer surface irregularities should be solved with the adequate tooling selection at the die, maybe higher die gaps. This problem depends on the resin selection and specific die output, as it was demonstrated in the present work.

CONCLUSIONS

It is confirmed that the extrusion of metallocene resins is very different compared to the LDPE processing due to the fact that these materials have an extended Newtonian plateau. Fortunately, the obtained operation windows are more attractive for the industry due to the high productivity levels.

The filler masterbatches act mainly as a processing aid due to the carrier selection (resins with a high melt index) and at the same time, a positive reinforcing effect can be achieved.

It appears further tailoring of the optimum screw design with a barrier section and different mixing devices but keeping under control the viscous dissipation problem. Processability improvements of these resins and their blends are required. A future contribution to this project will be to analyze the same blends using different screw configurations including novel mixing heads, and to compare the agreement between simulations and experimental results.

ACKNOWLEDGMENTS

All the measurements were performed at the Instituto de Capacitación e Investigación del Plástico y del Caucho (ICIPC in Medellín, Colombia. Thanks are due to DOW QUIMICA DE COLOMBIA S.A. who donated the mPE, COMAI S.A. for the master batch and QUIMICOPLASTICOS LTDA for the LLDPE.

REFERENCES

1   Knights, M., *Plastics Technology*, p.44, February 1995.
2   Butler, T., *Plastics Technology*, p.50, February 1995.
3   Cheng, C. Y., SPE ANTEC, p.2019, (1996).
4   Cheng, C. Y. and J. W. C. Kuo, SPE ANTEC, p.1942, (1997).
5   Domininghaus, H., **Plastics for Engineers**, *Carl Hanser Verlag*, Munich, Germany, (1993).
6   Khare A., C. Qin, M. T.K. Ling and L. Woo, SPE ANTEC, p.2198, (1998).
7   Rios A., T. Osswald, M. Noriega and O. Estrada, SPE ANTEC, p.262,(1998).

Processing Trends of Metallocene Linear Low Density Polyethylenes and Their Influence on Single Screw Design

K. R. Slusarz, C. A. Ronaghan, and J. P. Christiano
Davis-Standard Corporation, Technical Center Pawcatuck, Connecticut 06379, USA

INTRODUCTION

The changes in molecular architecture for these new metallocene LLDPE (m-LLDPE) have a dramatic impact on their physical properties. The physical properties of metallocene resins that are affected included the frictional, rheological, and melting behavior. These properties affect extrusion performance and screw design.[1,2] Specifically, it has been observed that the melting rate of m-LLDPE is strongly influenced by the material density and the amount of chain branching.[3] This study was undertaken to investigate the effect of these variables on the extrusion process and how they relate to screw design.

The first material was a conventional Ziegler-Natta (ZN) catalyzed LLDPE with a density of 0.920 g/cc and was used as a baseline for comparison. Two metallocene resins, which are produced using Constrained Geometry Catalyst Technology, and contain controlled long chain branching (LCB), were chosen. The two materials have densities of 0.908 and 0.902 g/cc respectively. These materials will be studied to see the effect of density on process performance. A second type of m-LLDPE produced on a single site catalyst such that it contains controlled short chain branching (SCB) with a density of 0.900 g/cc was selected to evaluate the effect of chain branching. The final comparison was with an identical m-LLDPE as previously described except that it contains a processing aid (PA) and antioxidant package. All of these resins have a melt index in the range of 1.0 to 1.2 g/cc and are representative of LLDPE's that are suitable for blown film applications.

Three different screws were evaluated to see the effect of screw geometry on the processing performance of these metallocene resins. These screws included two barrier designs, one of which is optimized for LLDPE and blown film, and a conventional Maddock mixing screw design with a 3:1 compression ratio.

EXPERIMENTAL

Extrusion tests were conducted on a 63.5 mm 24:1 L/D extruder with pressure transducers installed along the length of the barrel. The transducers were placed at 3 L/D intervals starting at 3.8 L/D from the feed opening. The melt temperature was recorded at the discharge of the extruder using an exposed junction melt thermocouple immersed to the center of the melt stream. All of the tests were conducted using a 41 cm wide film die with a die gap of 1.1 mm.

Three different screw designs were evaluated in this paper, two were barrier designs and one was a conventional mixing design. All three screws incorporated a Union Carbide mixing section in the meter section of the screw. The barrier screw had a secondary flight introduced in the transition section of the screw. This secondary flight had clearance greater than the main flight (barrier gap). This flight separated the solid bed from the melt pool, while the increased gap allows the newly melted polymer to pass from the solids channel to the melt channel. In both barrier designs the lead of the main flight within the barrier section was increased in order to maximize the available area for melting against the barrel wall. The optimized barrier screw for LLDPE had a barrier flight that maintained a constant lead throughout the barrier section that was greater than the lead of the main flight. This increased the volume for the melted polymer while decreasing the volume of the solids channel as you proceed down the barrier section of the screw. The barrier gaps and channel depths were selected to match the melting rate of LLDPE and the screw had an overall compression ratio of 1.92:1. In the second barrier design, the barrier flight not only had a greater lead than the main flight but it also changed its lead part way through the barrier section. This was done to optimize the amount of surface area available for melting at the beginning of the barrier section while allowing for sufficient volume of the molten polymer in the second half of the barrier section. This screw had an overall compression ratio of 2.52:1. The final design that was evaluated was a conventional square pitch 3.0:1 compression ratio mixing screw.

Five materials were evaluated in this study; one conventional LLDPE, four m-LLDPE. Physical properties for these resins are described in Table 1.

The test matrix included the five resins on an optimized LLDPE barrier screw. In addition, two different screw designs were evaluated to see the effect of screw geometry on the resins. The barrel profile was kept constant for all tests at: Zone #1 - 204, #2 - 218, #3 - 191, #4 - 177°C, while the die temperature was kept at 218°C. The breaker plate pressure was fixed at approximately 310 bar by a valved adapter. Overall performance was characterized by specific output, melt temperature, and power consumption. The solids conveying, melting, and pumping processes within the extruder were monitored and interpreted by means of pressure

Table 1. Materials properties

Material	MI, g/10 min	I_{10}/I_2	Density, g/cc	M.P., $^{\circ}$C	Chain branching
A	1.0	8.2	0.920	122	SCB
B	1.2	10.2	0.908	104	controlled LCB
C	1.0	8.7	0.902	100	controlled LCB
D	1.1	5.9	0.900	91	controlled SCB
E (D w/PA)	1.1	6.0	0.902	91	controlled SCB

traces which were recorded at very high sampling rates.[4] Extrusion output stability and thermal uniformity of the melt were determined by the total variation in pressure and melt temperature measured in the discharge adapter. The material was extruded into a film and examined visually for melt quality.

In addition to extrusion trials, screw "push-outs" were done on four of the resins to evaluate the melting characteristics of these resins. The optimized LLDPE barrier screw operating at 50 screw RPM, was used in all cases for the push-outs. The resins selected for this study were the conventional Z-N LLDPE (A), the m-LLDPE with a 0.902 g/cc density and LCB (C), the 0.900 g/cc m-LLDPE (D) and the corresponding m-LLDPE with PA (E). The resins were blended with 0.4% blue masterbatch, which was in a LLDPE carrier, in order to observe flow fields and melt pool generation. At steady-state, the extruder was abruptly stopped and 100% water cooling was initiated in order to freeze the material. After the extruder was fully cooled, the screw was extracted and the carcass was removed for evaluation.

RESULTS AND DISCUSSION

OPTIMIZED LLDPE BARRIER SCREW

Overall trends observed on each screw tested indicated an increase in specific output with the reduction of density. There was a trend of increasing melt temperature with a decrease in I_{10}/I_2. Although performance and pressure developments varied between the resins, process stability was acceptable for all the conditions tested. For the purpose of this paper, the 50 and 75 screw RPM conditions were closely studied.

The performance of each of the five resins on an optimized LLDPE blown film barrier screw was excellent. All five of the resins ran with total pressure variation under 0.74% and total melt temperature variation less than 0.88°C. Figure 1 shows the specific output for each material at both screw speeds. As reported in literature, the specific output increased with decreasing density. Resins C and D gave similar specific outputs which were within experimental error, indicating that the LCB did not have a strong influence on output rate in comparison to density.

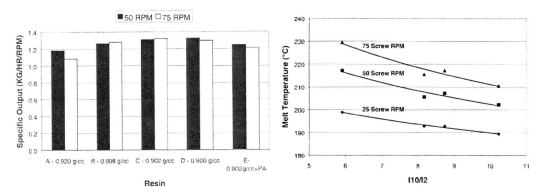

Figure 1. Specific output vs. resin and screw RPM - optimized LLDPE barrier screw.

Figure 2. Melt temperature vs. I_{10}/I_2 - optimized LLDPE barrier screw.

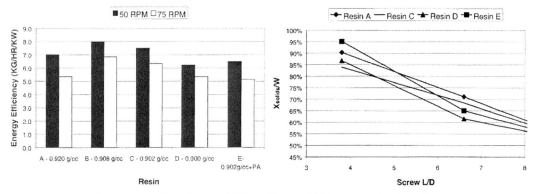

Figure 3. Energy efficiency vs. resin and screw RPM - optimized LLDPE barrier screw.

Figure 4. Melting progression down the screw.

As expected, the melt temperature increased with increasing screw RPM. Figure 2 shows the melt temperature versus the I_{10}/I_2.

The energy efficiency (KG/HR/KW) of LCB inLLDPE was consistently greater than the Z-N LLDPE and the SCB m-LLDPE (see Figure 3). This was due the shear thinning effect seen with the LCB m-LLDPE.

SCREW CARCASS ANALYSIS

The screw carcasses from the push-out experiments were cut at each pressure transducer to analyze the amount of melting occurring. Figure 4 shows the amount of solids (X_{solid}) vs. the width (W) of the channel at each pressure location. Figure 5 shows the pressure profile at each

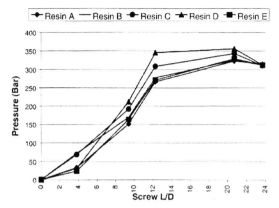

Figure 5. Nominal pressure profile along the screw - optimized LLDPE barrier screw.

pressure transducer for all the resins at 50 screw RPM. These figures indicate that resin C had the greatest amount of melt along with the highest pressures in the feed zone of the screw. This indicates the frictional coefficient along with the initial rate of melting were greater for resin C that of the conventional Z-N LLDPE.[6] The lower density m-LLDPE initiated melting earlier and at a higher rate than the Z-N LLDPE resulting in a greater specific output.

When comparing resin D to resin A, there was a similar trend. There was a greater initial amount of melt that is formed with resin D, which was consistent with lower density. In addition, resin D had a greater pressure profile along the screw yielding in a higher specific output.

When comparing resin C to resin D, resin C had a greater amount of melt and higher pressure initially. At P2, resin D had a greater amount of melt along with higher pressure. From these observations, one can infer that the LCB yields to greater frictional coefficients causing earlier melting, while the SCB narrow molecular weight inLLDPE melts much more rapidly once melting had initiated. Analysis of the cross channel pressure gradient for resin D, shows a much larger differential pressure, indicating a much higher melt viscosity resulting in a greater amount viscous energy dissipation once the melt film had established.

Figure 4 shows that resin E with PA delayed the start of melting. It also resulted in a lower pressure profile along the screw. This was most likely due to the interaction of PA on the screw and barrel surface. When the PA interacts at the polymer metal wall interface, it can decrease the shear stress transmitted at the wall. Once resin E initiated melting, its rate of melting was very high. This was observed at P2 where resin E had the second greatest amount of melt. Work has been published on the effect of PA that suggest that the PA coats the screw and barrel wall and change the interaction of the molten polymer with the metal resulting in reducing gels.[7] Our trials showed a significant improvement in the visual melt quality of the film, lower melt temperature, a slight reduction in specific output, and less cooling demand in barrel Zones #3 and #4.

ADDITIONAL SCREW DESIGNS

A second barrier screw with shallower channel geometry showed a similar trend with specific output versus density. Figure 6 shows the specific output of all the resins at 50 and 75 screw

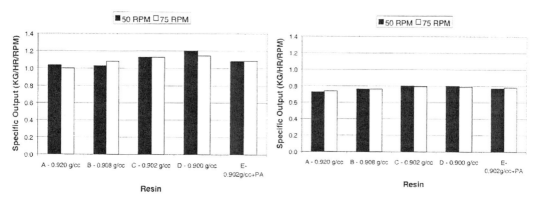

Figure 6. Specific output vs. resin and screw RPM - non-optimized LLDPE barrier screw.

Figure 7. Maddock mixing screw - specific output vs. resin and screw RPM.

RPM. As the figure indicates, the specific output of all the resins was lower than the optimized LLDPE barrier screw design. Melting instability was observed on this screw design. This was because the barrier design was not optimized for the high rate of melting that is typical of m-LLDPE. The melting instability compromises the integrity of the solid bed and upsets the steady state process. This is referred to as solid bed breakup (SBB).[5] Analysis of the pressure traces showed a repeatable steady-state upset that manifested itself as a spike in both the melt temperature and pressure measured in the adapter pipe.

The conventional Maddock mixing screw had the lowest specific output of all the screws. However., the trend of increased output rate with decreasing density was still apparent. Figure 7 shows the specific output as a function of material and screw speed. This screw design showed increased tendency for SBB along with greater spikes in the melt temperature and adapter pressure.

CONCLUSIONS

The results from this paper showed that changes in material properties due to recent advances in catalyst technology must be carefully considered in the design of single screws for metallocene polyethylenes. This study of m-LLDPEs showed increased melting rates and higher specific output rates. The optimized LLDPE barrier screw that was tested, processed both the LCB and SCB inLLDPE's successfully. However, further optimization is possible for the m-LLDPE based on the unique characteristics of each metallocene resin tested. Additional screw designs tested showed that instabilities due to the feeding and melting characteristics of the polymer can occur if the screw design cannot accommodate the high melting rates. As new polymers are produced with lower densities, high melting rates and high frictional coefficients, screw designs will have to change to accommodate these resin characteristics.

ACKNOWLEDGMENTS

We would like to thank Tom Butler of the Dow Chemical Company and Craig Erderly of Exxon Chemical Company for their contributions of materials and expertise.

REFERENCES

1   J. P. Christiano and K. R. Slusarz, TAPPI PLC Conference Tech. Papers, 483, (1998)
2   T. I. Butler, M. A. Spalding and A. Mayer, "A Processing Comparison of POP and EPE Polymers in Blown Film."
3   Y. S. Kim, C. I. Chung, S. Y. Lai, T. I. Butler and K. S. Hyun, SPE ANTEC, 42, 216, (1996).
4   C. D. Han, K. Y. Lee, and N. W. Wheeler, *Polym. Eng. Sci.*, **31**, 831, (1991).
5   R.T. Fenner, A. P. D. Cox, and D. P. Isherwood, *Polymer*, **20**, 733, (1979).
6   Y. S. Kim, C. I. Chung, S. Y. Lai, and K. S. Hyun, SPE ANTEC, 42, 207, (1996).
7   S. S. Woods and S. E. Amos, TAPPI PLC Conference Tech. Papers, 675, (1998)

Influence of mPE Grades on the Dynamic Properties of PP/mPE-Blends

Frank Raue and Gottfried W. Ehrenstein

Lehrstuhlf für Kunststofftechnik (LKT), University of Erlangen-Nuremberg, Am Weichselgarten 9, 91058 Erlangen, Germany

INTRODUCTION

The development of metallocene catalyst systems has led to numerous new materials. The new materials are built up of well-known monomer units but exhibit unique property profiles. The differences in the property profiles are attributed to the capability of the systems to allow the design of new molecular chains. An important group of polymers, which is synthesized with these catalysts, is metallocene polyethylene (mPE). This material has a very homogeneous copolymer distribution and a narrow molecular weight distribution. In particular, very low density mPE possesses elastomeric properties superior to commonly used PE grades.

In addition to the elastomeric behavior, mPE shows good dispersion in polypropylene (PP). Taking into consideration that PP exhibits low fracture toughness at low temperatures, it is often blended with an elastomeric material. Thus, it is being suggested to use mPE as a modifier for PP. In similar applications, mPE can also be used as a substitute for the ethylene propylene diene terpolymer (EPDM) in PP/EPDM-blends.[1,2] In our previous work, we have shown that some quantities like the tensile and dynamic behavior can be improved when substituting EPDM with mPE. This relationship can be pronounced if the soft phase is not crosslinked.[3] Low cost and a wide range of achievable mechanical properties favor compounding of the desired blend composition. For improved blend homogeneity, processing with twin extruders is preferred.

This work deals with the deformation behavior of PP/mPE-blends under static and dynamic loading. Dynamic load limits of the blend, which are necessary for several applications, are obtained using the hysteresis measurement method. Via this method the influence of the mPE grade on different mechanical properties is examined.

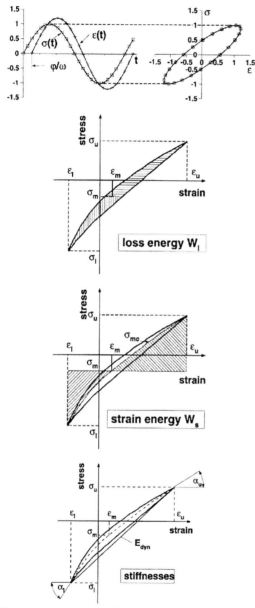

Figure 1. Construction of hysteresis loop and characteristic quantities derived from the hysteresis loop.

THE HYSTERESIS MEASUREMENT METHOD

The common S-N testing approach for material fatigue gives the number of cycles to failure for a given load. In contrast to this, the hysteresis method possesses two basic advantages when characterizing the dynamic material properties.[4] The first is an increase in information on the fatigue process, i.e., structural changes. The second advantage is a decrease in duration of the test. This becomes possible by evaluating the hysteresis-loop subjected to cyclic loading. The viscoelastic material behavior and material damage induce the loop in the stress-strain diagram as shown in Figure 1. Four different characteristic quantities; stress, strain, moduli and mechanical energy can describe the hysteresis loop itself. Basic properties shown in Figure 1 include the dynamic modulus E_{dyn} and the damping factor Λ:

$$E_{dyn} = (\sigma_u - \sigma_l)/(\varepsilon_u - \varepsilon_l) \quad [1]$$

$$\Lambda = W_l / W_s \quad [2]$$

Also the stiffness ratio

$$E_u / E_l = \tan\alpha_u / \tan\alpha_l \quad [3]$$

is very useful in the determination of the changes in the loop shape, i.e., for a thermoplastic elastomer.[5]

For the hysteresis measurement method, two different test patterns were performed. The first profile was the Stepwise

Load Increase Procedure (SLIP) which increases the dynamic load by a constant load ratio after a specified number of cycles. This profile is suitable for the rapid determination of load dependent characteristics. The second profile was used to determine the time dependent changes in characteristic quantities measured via the Single Load Level Procedure (SLLP) which uses a specified load amplitude. The load ratio R used for the SLIP analysis is defined as:

$$R = \sigma_l / \sigma_u \qquad\qquad [4]$$

EXPERIMENTAL

MATERIAL AND SAMPLE PREPARATION

For these examinations, a block copolymer polypropylene (Vestolen P9500, DSM) was used as the matrix material. With the use of a twin screw extruder (Leistritz LSM 30/34 GL, mass temperature: 200°C, rotational velocity: 100 rpm), the PP was compounded with four different metallocene ethylene-octene copolymer grades: Engage EG 8100, EG 8150, EG 8200 and EG 8842 by DuPont Dow Elastomers SA. Engage EG 8100 has a density of 0.868 g/cm³ (co-monomer content: 25 wt%). EG 8150 EG 8200 have a density only 0.002 g/cm³ greater than EG 8100 (24 wt% co-monomer), however these metallocene copolymers have a melt flow index at temperature of 190°C and mass of 2.16 kg of 1.0, 0.5, and, 5.0 g/l0 min, respectively. EG 8842 has a density of 0.857 g/cm³, a melt flow index of 1.0 g/l0 min, and is talcum powdered to reduce tacking of the pellets. Characteristic quantities of the materials as provided in the data sheets are given in Table 1.

Table 1. Quantities for the neat materials according to data sheets (MFI at 190°C, 2.16 kg, n.v.: no value)

Pure material	Shore A/D	σ_{max}, MPa	ε_{max}, %	MFI, g/10 min
P 9500	n.v.	32.0	>50	n.v.
EG 8150	75/20	11.9	750	0.5
EG 8100	75/22	10.3	800	1.0
EG 8200	75/21	7.6	>1000	5.0
EG 8842	n.v.	n.v.	>1000	1.0

After compounding the blends at a ratio of 60 wt% PP and 40 wt% mPE, the blend was molded into test specimens with the use of an Engel ES 200/50 HL at a mass temperature of 200°C, a mold temperature of 40°C, and a flow front velocity of 275 mm/min. The test specimens used for the experimental investigations had the sample geometry depicted in Figure 2.

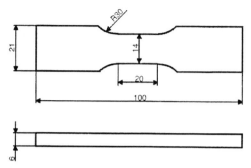

Figure 2. Testing geometry.

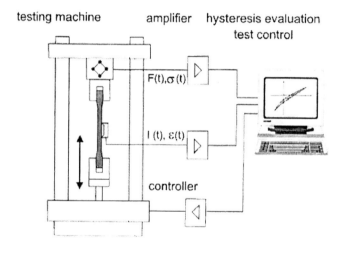

Figure 3. Test configuration.

TESTING EQUIPMENT AND PROCEDURES

In order to characterize the viscosity of the various blend components over the complete range of shear rates in the processing steps, the apparent shear viscosity was determined with a high pressure capillary rheometer (Verirel 7, Rosand, melt temperature: 200°C).

All mechanical tests were carried out at 23°C and 50 % relative humidity. Tensile tests for the static mechanical characterization of the material were run on a Zwick 1465 tensile testing machine. The crosshead speed during testing was 200 mm/min.

Tension-tension fatigue tests were performed on a Schenck servo-hydraulic testing machine. A load cell and an extensiometer, respectively, measured the force and longitudinal strains for the calculation of the hysteresis curve as shown in Figure 3.

The samples were subjected to stress controlled sinusoidal forced oscillation. The frequency in the tests was 0.35 Hz in order to avoid viscous heating of the samples. SLIP with a load ratio R=0.01 was conducted in order to obtain load limits for dynamic loading. The stress was kept constant during a period of 250 cycles and then stepped to the next higher level. SLIP show the validity of the resulting load limits.

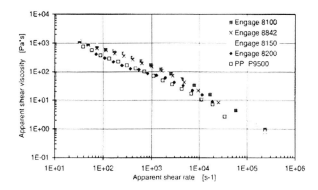

Figure 4. Viscosities of the pure materials at 200°C.

RESULTS AND DISCUSSION

VISCOSITY

Figure 4 shows the viscosity of the examined materials at 200°C. The viscosity of the different mPE's is comparable at higher shear rates, relevant to the shear rates obtained during the injection molding process. At lower shear rates, the high MFI grade EG 8200 exhibits the lowest viscosity. The PP has a lower MFI-value than the mPE's, i.e., higher viscosity at low shear rate, and is also slightly less viscous when high shear rates are applied. Since the phase with the lower viscosity usually forms the matrix in a copolymer system, PP would represent the major phase. However, at a viscosity ratio close to 1.0, an interpenetrating network may also be possible. Examinations to define the morphological states are yet to be investigated for the blends obtained under these processing conditions.

TENSILE TESTS

The stress-strain diagrams of the matrix material, the blends with 40 wt% soft phase, and the pure mPE grades are shown in Figure 5. As expected, the modulus and the tensile strength both decrease while the elongation at break is improved as the elastomer content increases. Investigating the results of the four blends further, it is evident that the blend partners with a low MFI EG 8150 and EG 8100 lead to a higher modulus and tensile strength than the other two

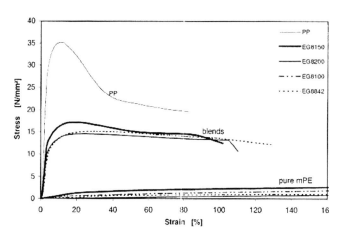

Figure 5. Stress-strain diagram of PP, mPE and their blends (23°C, 200 mm/min).

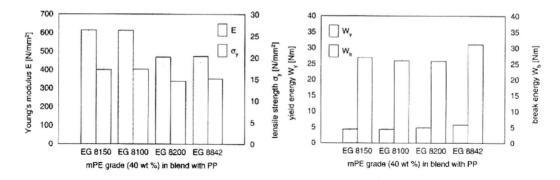

Figure 6. Young's modulus and tensile strength of PP/mPE blends (23°C, 200 mm/min).

Figure 7. Yield and break energy of PP/mPE blends (23°C, 200 mm/min).

mPE's, also shown in Figure 6. These results correlate well with the lower values for these pure materials.

A different result is obtained when calculating the yield and break energy by the area underneath the stress-strain curve up to the strain at yield stress or the break strain, respectively, as shown in Figure 7. Here, the better elongation behavior results in a considerably improved energy value for EG 8842, the material that contains talcum and has the lowest density, whereas the other blends do not differ much.

DYNAMIC EXPERIMENTS

Table 2. Quantities for PP/mPE blends with 60 wt% PP and 40 wt% mPE

Blend (PP+...)	E MPa	σ_{max} MPa	ε_{max} %	σ_Λ MPa
EG 8150	614	17.2	102	10.5
EG 8100	612	17.3	97	10.5
EG 8200	469	14.5	110	8.5
EG 8842	481	15.1	129	8.0

Figure 8 depicts characteristic quantities of the hysteresis measurements on specimens of the 60%PP/40%EG8100 and 60%PP/40%EG8200 blends, respectively (stress-strain diagrams show the upper, mean, and lower values). The dynamic modulus of the EG 8200 blend starts of at a lower level than the EG 8100 blend, which is in accordance with the tensile test results. The slope of the curves is similar, giving proof for a comparable deformation mechanism. However, differences do arise due to the different mechanical characteristic quantities. The EG 8200 blend shows a more dramatic decrease in the modulus and an increase in the

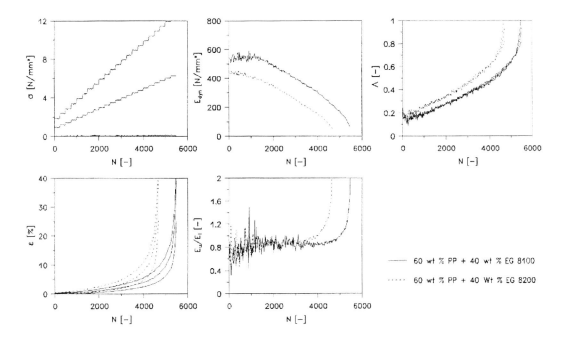

Figure 8. PP/mPE blends in SLIP (R=0.01, f=0.35 Hz).

damping, strain and the stiffness ratio of the upper to lower modulus of the hysteresis loop than the EG 8100 blend at low stress levels. Therefore, EG 8100 possesses a higher dynamic load limit. Table 2 lists the determined load levels at which dramatic changes in the damping, dynamic modulus, or stiffness ratio were observed.

CONCLUSIONS

PP modification was conducted in this research by the compounding of a PP with four different grades of mPE. Considering equal densities of the blends, the low viscosity grades showed a higher modulus and tensile strength, whereas at a similar viscosity, the grade with the lowest density gave the best energy absorption. The determination of the dynamic quantities using the hysteresis measurement method has shown load limits at which the values change dramatically. The influence of processing parameters on the mechanical behavior of PP/mPE-blends, especially at different frequencies and temperatures, as well as the morphology resulting from the processing conditions are the subject for further investigations.

ACKNOWLEDGMENTS

The financial support by the Deutsche Forschungsgemeinschaft (DFG) and the material support by DuPont Dow Elastomers, Geneva,. and DSM, Gelsenkirchen, is gratefully acknowledged.

REFERENCES

1   Karger-Kocsis, J. (Ed.), **Polypropylene - Structure, Blends and Composites**. *Chapmann & Hall*, London (1994).
2   Sylvest, R. T.; Lancester, G.; Betso, S. R., New Elastomers from Insite Technology, *Kautschuk, Gummi, Kunststoffe*, **50** (1997) 3, p. 186 - 191.
3   Raue, F.; Ehrenstein, G.W., The Static and Dynamic Behavior of PP/EPDM- and PP/POE-Blends using the Hysteresis Measurement Method, ANTEC, Atlanta (1998), p. 1442-1446.
4   Hoffmann, L.; Pavsek, V.; Schernme, M.; Dratschmidt, F.; Raue, F. Zysk, T.; Ehrenstein, G.W., Hysteresis Measurements - Making Fatigue Testing More Expressive. 1st Int. Conf. Mechanics of Time Dependent Materials, Ljubljana (1995), p. 17-22.
5   Raue, F.; Ehrenstein, G.W., Elastomermodifizierte Thermoplaste unter statischer und dynamischer Beanspruchung. Conf. Proced. Polymerwerkstoffe, Merseburg (1996), p. 85-89.

"Single-Site" Catalyzed Polyolefin for Fresh-Cut Produce Packaging. A Comparison Between Monoextruded Blends and Coextruded Film

V. Patel, S. Mehta, S. A. Orroth and S. P. McCarthy
Institute for Plastics Innovation, University of Massachusetts, Lowell, MA 01854

INTRODUCTION

Produce packaging demand has increased rapidly over the last two decades. The retail (consumer) market trend is more and more towards fresh-cut produce because of convenience, nutritional concerns, and higher spoilage of the fresh produce before consumption. After harvest, the quality of produce declines due to changes in color, texture, aroma, flavor and deterioration of vitamins, which makes them less acceptable to the consumer. The quality of produce is affected mainly by two factors, the chemical effect of atmospheric oxygen and growth of aerobic spoilage microorganisms. These factors are highly sensitive to temperature and can be slowed down considerably in cold storage but this is not sufficient to adequately extend the shelf life of produce. Thus, the requirement of modified atmosphere produce packaging is highly desirable to extend the shelf life of fresh produce. The most important criteria for good produce packaging are the selective permeability to specific gases such as oxygen, carbon dioxide, ethylene and water vapor at normal storage temperatures. Prior to the 1980s there was not a single film that met both the permeability and other performance requirements of fresh-cut produce packaging. Hence, artificial permeability of a film was achieved by incomplete sealing or by perforating the film. Gas and water transmission through the perforations is proportional to the effective hole diameter, the number of holes per unit film area and the uniformity of hole distribution over the package surface.[1]

The concept of atmosphere modification inside the package has developed rapidly over the last decade. The artificial atmosphere created inside the package slows down the metabolic processes in the produce and significantly prolongs self-life of the commodity. Film ventilation by perforation is not suitable for this type of packaging, because gas and water transmission properties vary and depend largely on the perforations and not on the permeability properties of the film. Simultaneously growth of packaged produce, which is not only pre-washed but also precut, is increasing tremendously. Fresh-cut produce (FCP) has larger surface area, which gives higher respiration rates than whole fresh produce. Hence the demand for films with higher oxygen transmission rates and lower moisture transmission rates is ever increasing.[2]

Modified atmosphere packaging (MAP) is a newly developed technique where the shelf life of pre-cut produce is extended by inhibiting chemical, enzymatic and microbial spoilage.[3] This allows preservation of the fresh state of the produce without the chemical treatments used for other shelf life extension processes. Traditionally, ethylene vinyl acetate copolymers (EVA) and ultra low-density polyethylene (ULDPE) have been used for these types of applications. However, new single-site catalyzed resins have replaced these resins because of some significant advantages such as excellent optical properties, very high oxygen transmission rate and outstanding seal performance. Typical product structure used by many film processors is one-component monolayer films; resin blends extruded into monolayer films, or coextruded films. Linear low-density polyethylene (LLDPE) and plastomer resins are used extensively in fresh-cut produce packaging applications. Each resin has its advantages and disadvantages in terms of processing, gas transmission characteristics and physical properties due to the difference in their molecular structure. However, comparative information on the processability, gas transmission rate and film physical behavior of these blends is still limited in the literature.

The purpose of this research is to compare the properties of monoextruded film of LLDPE/plastomer blends with coextruded film from the same materials at the same resin ratio. A new bimodal resin film will also be compared with these films for required properties of fresh-cut produce packaging.

EXPERIMENTAL

Plastomers (POP1 & POP2), bimodal resin, LLDPE resins and coextruded films used in this study were supplied by Dow Chemical Co. POP1 and POP2 have densities, 0.9308 & 0.8996 g/cc, slip additives, 2000 & 3500 ppm, antiblocks 3000 & 6000 ppm respectively. Bimodal & LLDPE resins have densities 0.9175 & 0.922 respectively. The monolayer (single layer) blown film was made by using sterling blown film extrusion line. The die (die diameter, 5"; die gap, 0.03") used in this study is a spiral mandrel die. All films were extruded at same temperature profile (149°C, zone 1; 171°C, zone 2; 216°C zone 3, & 205°C, zone 4). Adapter & die temperatures were kept at 205°C & 215°C respectively. Blow up ratio (BUR = 2) and screw speed (60 rpm) were also kept constant throughout processing. To obtain desired film thickness, varying takeoff speed varied the draw down ratio (DDR). Coextruded films were supplied by Dow Chemical Co., Freeport, TX for comparative study. The feed block type blown film die was used for coextrusion.

A DuPont TA 2000 with dual cell DSC-912 calorimeter was used to determine the thermal transitions of the polymer blends. The weight of the sample was about 6 mg. Scans were carried out from 0 to 160°C at a constant heating rate of 10°C /min. Tensile testing of the films was carried out according to ASTM D882 using an Instron 1137 tensile tester. Test was carried out at cross-head was 500 mm/min (20 in/min) and load cell was 200 pound. The standard deviation values for modulus were found to vary between 5 to 15%. Tear resistance testing was performed using an Elmendorf Tear tester (3200 gm) instrument, in accordance ASTM D1922. The standard deviation values for tear resistance were found to be within 4%. Impact resistance testing was performed using a Free-Falling Dart Impact Tester, in accordance with ASTM D1709 (Type-A).

Oxygen transmission rate (OTR), carbon dioxide transmission rate test (CO2TR) & water vapor transmission rate (WVTR) were carried out at Dow Chemical Co. OTR was performed using moist gases at ambient temperature on an Ox Tran 10/50 unit, in accordance with ASTM D3985-81. Oxygen concentration was 1% and data has been adjusted to 100% oxygen concentration values. CO2TR was done on Mocon MultiTRan unit, using a Dow Test procedure. OTR & CO2TR values are reported in cubic centimeters at standard temperature and pressure (STP = 0°C and 1 atm) and normalized for thickness. WVTR test was performed using Mocon Permatran W unit according to ASTM F1249-90. Heat seal strength & hot tack strength tests were carried out at Dow Chemical Co. using Dow test procedure. Heat sealing strength was performed by mounting one-inch (25.4 mm) wide strips on a Topwave Hot Tack Tester. Dwell time was 0.5 s and seal pressure was 40 psi (0.275 MPa). Sealed samples were stored at ASTM conditions (22°C, 50%RH) for 24 hrs and then tested on an Instron Tensiometer at a crosshead speed of 2 in/min (50 mm/min). Hot tack strength was performed at the same Topwave tester with seal time of 0.5 s, delay time of 0.2 s, peel speed of 150 mm/s, and seal pressure of 40 psi.

RESULTS AND DISCUSSION

The film obtained by the blown film process is known to be influenced by many variables. It is interesting to note that at equal melt flow index (1 g/10 min), the polyolefin plastomers (POPs) produced via INSITE technology are easier to process (lower back pressure and motor amps) than a conventional linear low density polyethylene (LLDPE). Generally, as molecular weight distribution (MWD) becomes narrower, materials behave more like a Newtonian fluid and become less shear sensitive. However, POPs with the incorporation of long chain branching (LCB) exhibit improved processability. Improved processing is defined as higher output rates at a given shear stress, less susceptibility to melt fracture, and sufficient melt strength for good bubble stability during blown film fabrication.

DSC scans of selected film samples indicate that the POP2 has lower melt temperature (98°C) and heat of fusion (50 J/g) than POP1 (102°C & 70 J/g). This results from higher comonomer concentration, which controls the amount of short chain branches, the length of which depends on comonomer type, and thus the resin density. The higher the comonomer concentration, lower the crystallinity, which lowers the density of resin. The LLDPE film shows both higher melting temperature (124°C) and heat of fusion (106 J/g) compared to POPs.

This is due to high crystalline structure. DSC scans of monoextruded film of the POP1/LLDPE (40/60 wt%) blend and the coextruded film show that the components retained their identity in the coextruded film, with little co-crystallization and the endothermal peaks did not overlap. This shows that the POP1 & LLDPE crystallize separately. On the other hand, the monoextruded blend shows intimate mixing of both materials. Martinez[4] observed similar results for cooling curves for a 50/50 LDPE/LLDPE blend. The bimodal resin film shows the highest crystalline melting temperature and a much broader peak as compared to the other three pure materials. This indicates that the bimodal resin might require different processing conditions than were used in this study.

The tensile modulus (stiffness) and tear resistance in both machine direction (MD) and transverse direction (TD) increases with increasing percentage of LLDPE for both POPs. Figures 1 and 2 show the comparison between monoextruded blends and coextruded film of 60 wt% of LLDPE with bimodal resin film for tensile modulus and tear resistance respectively. Most of the films show higher tensile modulus and tear resistance in TD than in MD. Figure 1 shows that the modulus of monoextruded and coextruded film of POP1 and bimodal resin film is almost equal and higher than that of monoextruded film of POP2. This was because stiffness is related to film density (crystallinity). DSC results show that the POP2/LLDPE blend film has a little lower heat of fusion in comparison to the POP1/LLDPE blend, however it has broader melting transition. It could be due to inhomogeneity of the blend caused by comonomer concentration and distribution that might affect the crystalizability of the components in the blend and finally the mechanical properties. For most of the cases as draw dawn ratio (DDR) increases, modulus in the MD increases. This is because as DDR increases, amorphous phase orientation and chain extension along MD increases. Tensile modulus in TD remained almost constant as blow up ratio (BUR) remained constant. Figure 2 demonstrates that monoextruded blend films have a little higher tear resistance than coextruded film.

Figure 3 shows that both the POP/LLDPE blend films did not fail at the 1050 g test maximum load limit up to 60 wt% LLDPE. This indicates that both POPs provides excellent toughness. However, for 80 and 100 wt% LLDPE impact resistance decreases. This is because as density (crystallinity) increases the film becomes more brittle. Thus, fracture

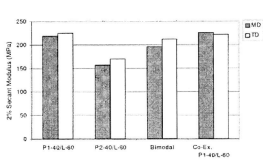

Figure 1. Comparison between monoextruded blends and coextruded film of equal content of LLDPE (60 wt%) with bimodal resin film for tensile modulus.

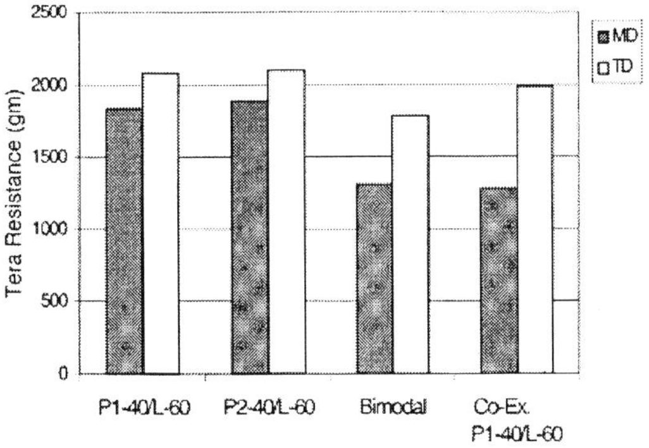

Figure 2. Comparison between monoextruded blends and coextruded film of equal content of LLDPE (60 wt%) with bimodal resin film for tear resistance.

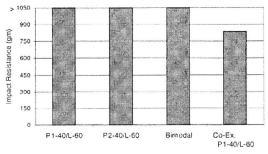

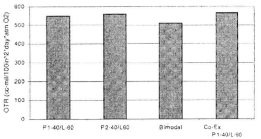

Figure 3. Comparison between monoextruded blends and coextruded film of equal content of LLDPE (60 wt%) with bimodal resin film for impact resistance.

Figure 4. Comparison between monoextruded blends and coextruded film of equal content of LLDPE (60 wt%) with bimodal resin film for OTR.

resistance decreases and so does impact energy. The impact resistance of monoextruded blend and bimodal films is higher than the coextruded film. It is not possible to explain the effect of film thickness on impact resistance because maximum capacity of the test equipment is 1050 g. However, it can be seen that all films had excellent impact resistance, and have higher values than the requirement of fresh-cut produce packaging.

Figures 4, 5 and 6 show comparison between monoextruded blends and coextruded film of 60 wt% of LLDPE with bimodal resin film for oxygen transmission rate (OTR), carbon dioxide transmission rate (CO_2TR) and water vapor transmission rate (WVTR) respectively.

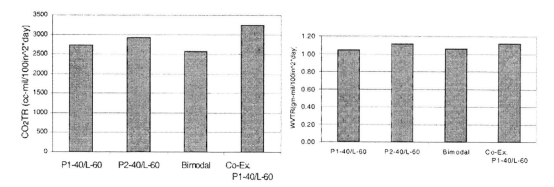

Figure 5. Comparison between monoextruded blends and coextruded film of equal content of LLDPE (60 wt%) with bimodal resin film for CO2TR.

Figure 6. Comparison between monoextruded blends and coextruded film of equal content of LLDPE (60 wt%) with bimodal resin film for WVTR.

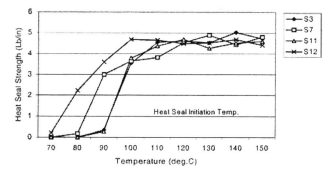

Figure 7. Comparison between monoextruded blends and coextruded film of equal content of LLDPE (60 wt%) with bimodal resin film for heat sealing strength. Codes: S3:P1-40/L-60; S7:P2-40/L-60; S11:bimodal; S12:Co-ex. P1-40/L-60.

As LLDPE contents increases permeability of the film decreases for both POPs. This is because permeability requires free volume, so gases permeating through the polymer are limited to the amorphous phase in semicrystalline polymers. Diffusion occurs mainly in the amorphous region, and motion around the crystalline area is a tortuous and difficult path.[5] Permeability depends upon both solubility and diffusion; since crystallinity reduces both, it has even a greater effect in reducing their product. Thus, as LLDPE content increases, total crystallinity of the film increases, which reduces the permeability of the film. It can be noted that all film samples has almost equal OTR and WVTR values even though some film samples have different crystallinity. This was unexpected and might be the result of crystalline alignment/orientation, which could alter permeability and affect the pathway available for diffusion.[6] Orientation of the amorphous phase also decreases permeability. The CO2TR for the coextruded film was higher than for POP2 blend with 60 wt% LLDPE, which was unexpected and might be because of differences in processing conditions. Generally, permeability of the film decreases with increase in orientation (DDR) of the amorphous region as well as the crystalline region that creates a more tortuous pathway. Processing conditions (e.g. cooling, DDR) can cause variation in film permeability as a result of degree of orientation.

Figure 7 shows comparison between monoextruded blends and coextruded film of 60 wt% of LLDPE with bimodal resin film for heat seal strength. Superior sealing strength is achieved, when two polymer surfaces are pressed together under adequate pressure for sufficient time so that the polymer chains can diffuse and intermix across the surface boundary. As LLDPE content increases, heat seal initiation temperature also increases. This is expected because as LLDPE content increases, crystallinity and melting point of the film increases, which reduces the mobility of segments of the adjacent amorphous polymer phase and thus increases the heat seat temperature at which they have enough mobility to produce an adequate heat seal. Also POP2 blends have lower heat sealing temperatures than POP1 for equal LLDPE content because of lower overall crystallinity. Coextruded film shows lower heat sealing temperature and higher hot tack strength over a wide sealing range compared to other films. This is an important advantage of coextruded film over monoextruded film. In case of the coextruded film 100% POP1 (narrow melting point range and a lower melting point), both heat seal layers maximized the rate of polymer diffusion and chain entanglement between the layers. Heat seal initiation temperature of all films is the same even though film thickness varied. This may be due to the fact that dwell time and pressure applied are too high at the time of seating and could not differentiate the values for variable film thickness. It is observed that ultimate heat seal strength increases as the film thickness increases. This was because heat seal strength (lb/in) and hot tack strength (N/in) were normalized by width of the films and not by the thickness. All films exhibit similar ultimate hot tack values, ranging only from 2.5-3.0 N/in.

CONCLUSIONS

The combination of narrow MWD and CD with LCB of the POPs is shown to have better processability than a similar melt index LLDPE polymer. Within a given family of polyolefins, crystallinity generally increases linearly as density increases. Crystallinity increases stiffness at the expense of toughness, permeability and sealability. Controlling crystallinity and degree and extent of crystalline orientation can alter these properties. DSC scans confirm that each material (POP1/LLDPE) retains its identity in the coextruded film, while blended film shows an intimate mixing of both materials. Bimodal resin film shows high crystallinity, which might affect its flexibility performance.

The monoextruded blend of POP1 40 wt%/ LLDPE 60 wt%, shows somewhat better mechanical properties than the coextruded film of same polymers at the same resin ratio, except for CO2TR. The bimodal resin film presents a good option, where superior mechanical properties are required. Permeability of the monoextruded blend and the coextruded film of the same materials at the same resin ratio appears to be the same. Bimodal resin film with moderate permeability could be used for low respiration rate fresh-cut produce. Overall, all films had their distinct advantages and disadvantages. The permeability of the package must be carefully matched to the respiration rate of the fresh-cut produce inside the package. For moderate to low respiration fresh-cut produce, coextruded film would be better than monoextruded films because of excellent sealing performance. In monoextruded film, bimodal resin film could be more useful than POP blends for low respiration rate fresh-cut produce. For high respiration requirements, coextrusion of a thin layer of LLDPE (provides modulus) with a thick layer of POP (provides OTR) is an alternative solution.

ACKNOWLEDGMENT

The authors would like to thank Dr. Jeffrey Wooster and the Dow Chemical Co. for supplying materials & other support.

REFERENCES

1   Peleg, K., **Produce Handling Packaging & Distribution**, *AVI Publishing Co., Inc.*, CT, 1985.
2   Wooster, J., *J. Plastic Film Sheeting*, **14**, Jan. 1998, pp.77-89.
3   Farber, J., and Dodds, K., **Principles of Modified Atmosphere and Sous Vide product Packaging**, *Technomic Publishing Co., Inc.*, PA, 1995.
4   Martinez, F., Mazuera, G., and Parikh, D., *J. Plastic Film Sheeting*, **6**, 1990,44.
5   Deanin, R., **Polymer Structure Properties & Applications**, *Canhers*, MA 1972.
6   Krohn, J., et.al., *J. Plastic Film Sheeting*, **13**, 1997, 327-335.

Rheology and Processing Experience of New Metallocene Polyolefins

Atul Khare, Chuan Qin, Michael T. K. Ling, and L. Woo

Baxter International, Inc. Round Lake, IL 60073

INTRODUCTION

The so called metallocene revolution has created new families of polyolefins with unprecedented properties. Very high comonomer contents in polyethylenes gives low modulus, high flexibility and some cases, thermoplastic elastomeric properties. The ability to incorporate short chain branches homogeneously have created polyethylenes of excellent optical clarity. This property is very important in medical and food packaging.[1] However, due to the high productivity and homogeneous short chain branching distribution and the very narrow molecular weight distribution, very Newtonian flow curves[2] are quite common for these polymers. A rheological comparison on these polyolefins will be presented and related to actual large scale processing experiences.

EXPERIMENTAL

Linear viscoelastic studies in the melt were conducted using either a Rheometrics Fluid Rheometer (RFR) or the Advanced Rheology Expansion System (ARES) in the dynamic mode with 2.5 cm diameter parallel plate fixtures. Sample thickness of between 1.5 to 2 mm was used. Frequency range of between 0.1 and 300 radians sec^{-1} was covered with strain amplitudes of between 10 and 25%, depending on the viscosity of samples. In most cases, before data were taken, a strain amplitude sweep at 1 Hz was conducted between 1% and 35% to ensure that linear response were observed. In selected cases, capillary flow rheometry was also carried out by an Instron capillary rheometer with a capillary of 25:1 length to diameter ratio.

Processibility studies were conducted on a commercial scale Davids Standard 3.5" single screw extrusion line with a barrier screw of 24:1 L/D ratio and tubing fixtures to produce tubing of about 60 mm OD and 800 μm wall thickness.

Table1. Sample designation

Sample	Density	MI	Comonomer	Process
ss ULDPE	0.88	3.6	butene	solution
ss ULDPE	0.87	1.0	octene	solution
ZN VLDPE	0.89	0.5	butene	gas phase
ss ULDPE	0.885	1.0	octene	solution
LDPE	0.92	10	-	high pressure
EVA (18% VA)	0.92	0.45	vinyl acetate	high pressure
EVA (28% VA)	0.94	5	vinyl acetate	high pressure

Samples studied (Table 1) in this investigation included single site (both metallocene and non-metallocene) ultra low density polyethylenes (ULDPE) and very low density polyethylenes (VLDPE), low density (LDPE) as well as ethylene vinyl acetate copolymers (EVA) from the high pressure process.

RESULTS AND DISCUSSION

STRUCTURAL PARAMETERS AND PRODUCT/PROCESSING PERFORMANCE

One of the hallmarks of the single site catalyst is the highly active and controlled reactivity for incorporation of comonomers, resulting in high comonomer contents and homogeneous placement of short branches. The so-called short chain branching distribution (SCBD)[3] is very narrow. As a result, during the crystallization from the melt, very small and well-controlled crystallites are distributed throughout the material. This is in stark contrast to the Ziegler Natta multi-active site catalyst, where a multitude of comonomer incorporation are distributed throughout the molecular weight ranges. Typically, high molecular weight, relatively short branch free fractions are mixed with low molecular weight highly branched species. Hence, during crystallization, a broad distribution of crystallite sizes are produced. Since the high molecular weight, higher crystallinity (less branched species) crystallize first, they produce relatively large crystallites. However, since the light scattering power for micron and submicron particles varies as the 3rd power of particle diameter,[4] high optical haze results from these heterogeneous polymers. For example, for homogeneous polyethylenes, the optical haze as measured by ASTM D1003 showed a linear dependence with crystallinity, while a heterogeneous Z-N polyethylene with equal crystallinity showed at least an order of magnitude higher haze (Figure 1).

Another important parameter affecting the product and processing performance is the melting point and crystallization point. As a rule, the melting point marks the absolute highest temperature that the material can withstand before the onset of fluidity. And the crystalliza-

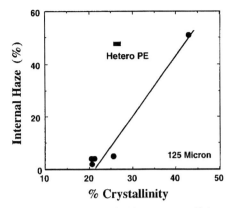

Figure 1. Homogeneous PE optical haze vs. crystallinity.

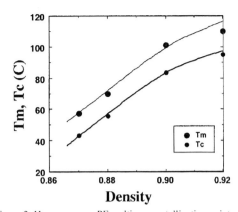

Figure 2. Homogeneous PE melting, crystallization point.

Figure 3. Metallocene syndiotactic polypropylene.

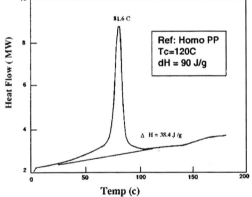

Figure 4. Syndiotactic PP crystallization.

tion point marks the minimum temperature that a fabricated article- either by extrusion or injection molding, can begin to assume its solid shape. In Figure 2, one sees that the homogenous polyethylene has a monotonic dependence of melting and crystallization points with crystallinity. It is quite obvious, that some of the low crystallinity polyethylenes pose serious challenges for heat removal during the fabrication process. Likewise, the metallocene produced syndiotactic polypropylene (Figure 3) despite the excellent optical clarity and toughness, exhibits a crystallization temperature of nearly 80°C (Figure 4) some 40°C lower than conventional isotactic polypropylenes.

RHEOLOGY DEPENDENCE ON STRUCTURE

Yet another characteristic of the single site polyethylene is the narrow molecular weight distribution (MWD) of around 2. This results in a rather Newtonian flow behavior toward varying shear rates or frequency. In other words, the viscosity versus shear rate curve is rather flat.

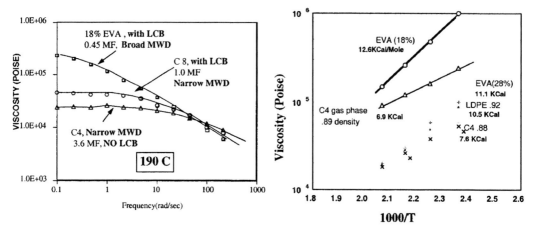

Figure 5. Rheology comparison.   Figure 6. Flow activation energies.

For broad MWD materials, the low shear or shear independent viscosity h_0 depends on the weight average molecular weight M_w for equal M_n. Hence, a rather high h_0 can be obtained. While at high shear rates, the low end of the molecular weight distribution becomes dominant resulting in a reduction in the apparent viscosity - hence a highly non-Newtonian behavior. Although the Newtonian behavior is desirable for controlling the melt relaxation times, important for processes like fiber spinning, for melt extrusion and injection molding the reduced shear thinning leads to excessive back pressure and reduced production rates. In addition, the excessively high shear stresses at production could also lead to melt fracture and other flow instabilities, causing drastic reductions in production speed.

Certain single site polyethylene, due to their catalyst's steric control and high activity, can contain low levels of long chain branching (LCB). The existence of long chain branching can benefit the flow behavior in two ways:

1   LCB tends to make the molecule more compact for equal molecular weights, and hence more shear thinning at higher shear rates.

2   In an analogous situation with high pressure polymerized low density PE (LDPE), a greater sensitivity of viscosity toward temperature is seen. This higher temperature dependence as measured by Arrhenius activation energy, can make the melt more "forgiving" during fabrication.

In Figure 5, for clarity we compare a selected subset of our samples in terms of shear thinning behavior. It can be clearly seen that a narrow MWD 3.6 MI single site PE in the absence of LCB was very Newtonian indeed. In comparison, a 1.0 MI, still narrow MWD single site PE exhibited higher Newtonian viscosity proportional to the MI, but crossover the 3.6 MI

Table 2. Activation energies

Material	LCB	ΔH, kcal/mole
LDPE	Y	10.8
18% EVA	Y	12.6
28% EVA	Y	11.1
0.88 C4Sol.	N	7.5
0.89 C4Gas	N	6.9
0.87 C8Sol.	Y	9.8

Figure 7 Tygon^{TM} tubing surface.

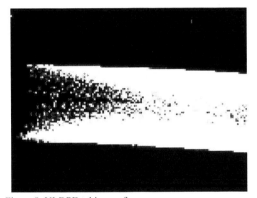

Figure 8. ULDPE tubing surface.

line to achieve lower viscosity at higher frequencies. However, the most dramatic difference lies with a high pressure polymerized EVA with 0.45 MI. The presence of both LCB and broad MWD allowed a much higher low shear viscosity, or in processing terms, "melt strength," or "melt tension," while maintaining a high shear viscosity equal or even lower than the two previous examples. The high low shear viscosity and the shear sensitivity frequently results in a strain hardening behavior when the elongational viscosity is measured, leading to much greater bubble stability in blown film and better parison strength in extrusion blow molding.

Figure 6 shows the Arrhenius plot for several of the samples. It is generally seen that long chain branched (LCB) samples exhibiting higher activation energies. Table 2 tabulates the measured activation energies.

MELT INSTABILITIES DURING PROCESSING:

Earlier, we stated that narrow MWD long chain branched polyethylenes possesses shear thinning behavior intermediate between the narrow distribution without LCB and the high pressure counter parts with both broad MWD and LCB. In a large scale processing trial, we compared the maximum output rates (linear line speed) of a tubing extrusion line up to the onset of melt instability. For comparison purposes, a commercial PVC Tygon^{TM} tubing's surface

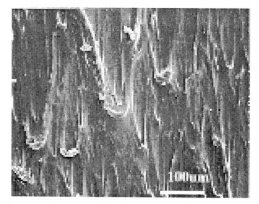

Figure 9. High rate ULDPE surface.

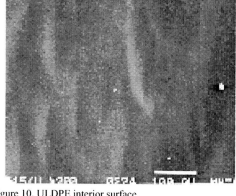

Figure 10. ULDPE interior surface.

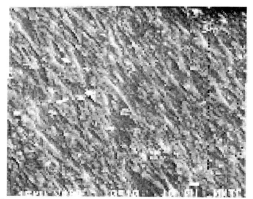

Figure 11. Textured PVC tubing exterior surface.

morphology was used (Figure 7). At equal production line speed of flexible PVC, the 0.885 density 1 MI resin exhibited visible outer surface irregularities (Figures 8 and 9). Fortunately, the interior surface, due to the somewhat reduced shear rates, showed a much smoother appearance (Figure10). This combination of considerable external roughness and smooth interior surface qualifies the tubing from a production rate consideration, although marginally. An additional comment is during typical medical tubing production, special surface texturing is applied to roughen the surface for better handling (Figure 11). Hence, the ULDPE rough surface caused by melt instability is quite acceptable, so long the overall tubing dimension can be maintained.

CONCLUSIONS

A broad based rheology and processing study was conducted on a series of single site and metallocene polyethylenes using high pressure LDPE and EVA as references. Confirming to what had been widely reported, the single site polymers with narrow molecular weight distribution and very Newtonian flow curves were significantly more difficult to process. Dramatic lowering of the melting and crystallization points also posed heat removal challenges. Incorporation of long chain branching (LCB) increased shear thinning and the Arrhenius activation energy of flow. Both factors helped to improve processibility sufficiently to be conditionally acceptable, but still inferior to that of the high pressure polymers like LDPE and EVA. It appears further tailoring of the molecular weight distribution would be required for processibility improvements.

REFERENCES

1   L. Woo, T. K. Ling, S.P. Westphal, Proc. Polyolefins VIII Intl. Conf. (Houston), 242, (1993).
2   S. H. Wasserman, Soc. Plas Eng. ANTEC Proc. 1129, (1997).
3   L. Wild, T. R. Ryle, D. C. Knobelloch, I. R. Peat, *J. Polym. Sci. Phys.*, **20**, 441, (1982).
4   C. B. Bucknall, **Toughened Plastics**, *Applied Science*, London, (1977).

Properties and Film Applications of Metallocene-Based Isotactic Polypropylenes

Aiko Hanyu and Rusty Wheat
Fina Oil and Chemical Company, Deer Park, TX

INTRODUCTION

Polypropylene metallocene technology allows molecular engineering that can control MW/MWD (molecular weight/MW distribution), various CC/CCD (comonomer composition/CC distribution), stereochemistry and stereoregularity.[1] All of these aspects of molecular structure control crystallizability, degree of crystallinity and morphology, which in turn influence processability and define final properties of end-use products. Because most of these molecular parameters do vary simultaneously with the structure of the metallocene catalyst system,[2] it becomes critical to understand the general relations among catalyst structure, polymerization mechanism, molecular structure, morphology, and processing to design ultimate end use properties.

Polypropylenes produced by the metallocene technology are expected to provide competitive properties over the conventional Z-N product.[1] Fina recently presented[3] high-MW miPPs which showed resin characteristics unique to the metallocene product, including low xylene solubles level, low MWD, low DSC melting point, etc. In the BOPP film application, the miPP oriented film demonstrated enhanced properties in terms of stiffness, COF, and permeabilities to water vapor and oxygen, as well as satisfactory optical properties. This paper presents more results from additional miPP products in both the BOPP and cast film applications.

Highly isotactic polypropylenes are a result of the predominant head-to-tail addition (1-2 insertion) of propylene monomer units, where methyl groups always have the same configuration with respect to the polymer backbone. A disruption of the long isotactic sequences creates a stereo defect, which lowers the isotacticity. The 2-1 insertion is a type of stereo defects resulting of the head-to-head insertion of propylene monomers, which leads to a

Figure 1. Schematic presentation of 2-1 insertion defect in isotactic propylene homopolymer.[2]

sequence of two adjacent methyl groups and two CH_2 groups shown in Figure 1.[2] The 2-1 insertion defects are present only in certain miPPs, but not in Z-N iPPs.[3]

The significant melting point depression observed in the miPPs[3] might be inherent to the distribution of the 2-1 insertion defect among chains.[2,4] The occasional 2-1 insertion defects may behave somewhat in the manner of random ethylene propylene copolymers where ethylene units in the chain function as the main defects. As for copolymer,[5] it may be understood that the decrease in melting point results from a distortion, caused by a defect, of the propylene helix and that this distortion is excluded from the lamellar crystallite, which consequently reduces the thickness of the lamella and thus lowers the melting point. Since the mechanical and thermal properties of the final products are closely related to the molecular structural homogeneity of the resins, it is valuable to characterize the defects sequence distribution.[6] This is attempted in this work using a segregation fractionation technique based on a stepwise isothermal crystallization by DSC[7] to make a comparable study of the heterogeneity in defect unit distribution between the standard Z-N and metallocene iPPs.

Table 1. Sample characterization of Z-N and metallocene iPP products

Sample	STD-3	HCP-2	MET-3	MET-6	METC-5
MFR[a]	2.7	1.9	3.3	5.5	5.0
XS[b]	4.5	0.7	0.2	0.3	0.7
C_2[c]	0	0	0	0	0.5
mmmm[d]	94.7	97.3	96.3	94.0	-
racemic[d]	1.5	0.2	0.7	1.5	-
2-1 ins.[d]	0	0	1.0	0.6	-
ZSV[e]	6.1	9.9	4.1	3.2	2.9
J_{or}[f]	4.4	3.6	1.6	1.7	1.3
M_w/M_n[g]	6.5	7.3	3.0	3.9	3.3
T_{m2}[h]	160	163	151	150	148
T_c[h]	107	111	108	109	107

Notes: a - MFR in g/10 min, b - xylene solubles in wt%, c - ethylene in wt%, d - by ^{13}C NMR for the xylene insolubles fraction in mol%, e - zero shear viscosity by rheometry @ 210°C in x10³ Pa-s, f - recovery compliance in x10⁻⁴ Pa⁻¹, g by GPC, h - DSC melting and recrystallization peaks in °C with heat/cool rates of 10°C/min.

EXPERIMENTAL

MATERIALS

The resin characteristics of the representative samples employed here are summarized in Table 1. The sample labels STD and HCP stand for standard and high crystallinity polypropylene produced by Z-N catalyses, respectively. The MET stands for miPPs and METC for metallocene copolymers. The suffix denotes melt flow rate (MFR) of a specific sample. All the resins were produced on the Fina's commercial lines.

The miPPs exhibit characteristics, such as low xylene solubles level, narrow MWD, low recovery compliance (J_{0r}), presence of 2-1 insertion defect, and low DSC melting point.

DSC AND SIST

All the DSC runs were obtained using a Perkin Elmer DSC-7 apparatus under nitrogen. Two DSC procedures were employed. Method A was a common one, where a weighed sample was melted at 210°C for 5 min and then cooled at 10°C/min to 50°C, subsequently heated at 20°C/min up to 190°C to obtain the melting endotherm. Method B was a stepwise isothermal segregation technique (SIST),[6-8] where a weighed sample was first melted at 210°C for 20 min and then quenched to the highest temperature to be crystallized isothermally. Subsequently, the sample was subjected to isothermal crystallization successively at intervals of 5°C over its melting range. The time for each step of isothermal crystallization was about 9-10 h. In the present set of experiments the temperatures of isothermal crystallization were 155-135°C and 100°C for 0.5 h for Sample MET-3, and 175-135°C for Sample STD-2. At last, the cooled sample was heated at 20°C/min and the endotherms were recorded as for Method A. The endotherms were normalized by sample weights.

LABORATORY BOPP FILM PROCESS

The sample resins were first extruded on a 1.25" (3.2 cm) extruder and casted into sheets of 4.1 mm. The standard extrusion parameters were modified for the miPPs by lowering the melt temperature and screw speed by approximately 18°C and 17%, respectively. Subsequently, simultaneous-biaxial orientations were carried out on a laboratory scale film stretcher, TMLong®. Each sheet was normally stretched to four times of its original dimension at a constant rate of 5 cm/s. The draw temperature was varied. Outputs of force vs. time (or stress vs. strain) data were collected and processed for analyses.

BOPP FILM PROCESS

The BOPP film processing was conducted on a 60" (1.5 m) continuous pilot tenter line. The sample resins were subjected to biaxial orientation procedures in two sequential steps. The draw ratio was five first in the machine direction (MD) and then eight in the transverse direction (TD) to produce a monolayer film of 18 μm. Properties of the BOPP films were evaluated according to ASTM procedures and listed in Table 2.

CAST FILM PROCESS

The miPP resins were extruded on a 3.5" (8.9 cm), 32:1 L/D extruder and cast into films of 0.05 mm at approximately 215°C melt, 260°C die and 16°C cast roll temperatures, and at 30.5 m/min line speed. The melt temperature of the standard iPP resins was kept at 230°C. Properties of the cast films were evaluated according to ASTM procedures and listed in Table 3.

Table 2. BOPP film properties of Z-N and metallocene iPP products

Sample	STD-2	MET-3	MET-6	METC-5
MFR	1.8	3.3	5.5	5.0
Haze[a]	0.3	0.9	0.2	0.2
Gloss[a]	99	97	98	94
MVTR[b]	3.4	2.6	2.9	3.2
OTR[c]	2600	2300	2900	2700
COF[d]	0.7	0.4	0.3	0.3
MD				
Modulus[e]	1.3	1.5	1.7	1.9
Tensile[f]	130	160	140	150
Elongation[a]	140	120	160	170
Shrinkage[a]	7	11	4	4
TD				
Modulus[e]	1.8	2.3	2.7	2.8
Tensile[f]	300	300	250	260
Elongation[a]	50	40	50	60
Shrinkage[a]	10	21	10	8

Notes: a - units in %, b - in $g/m^2/day$, in $cc/m^2/day$, d - kinetic COF, e - 1% secant modulus in GPa, f - tensile strength @ break in MPa

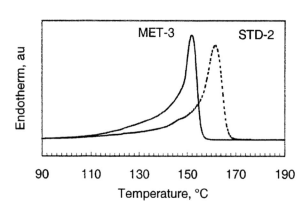

Figure 2. DSC second melting behavior for Z-N and miPP samples by Method A heated at 20°C/min.

RESULTS AND DISCUSSION

Figure 2 compares the DSC thermograms obtained after Method A for the Z-N and miPP resins. Both resins show a similar asymmetric broadening (peak tail) of the melting peak in the low-temperature range. The peak maximum of MET-3 is 10°C lower, although it has a comparable isotacticity to STD-2 in terms of meso pentad (%mmmm). The half-width of the miPP is narrower than that of the Z-N resin. The significant melting point depression, mentioned previously in the

Table 3. Cast film properties of Z-N and metallocene iPP products

Sample	STD-8	MET-6	MET-9	MET-14	METC-5
MFR	7.9	5.5	9.4	13.7	5.0
Haze[a]	4.6	2.0	1.2	0.4	1.8
Gloss[a]	59	75	79	81	77
Impact[b]	440	460	410	440	440
MD					
Modulus[c]	480	520	500	560	480
Tensile[d]	46	44	36	40	47
Elongation[a]	760	710	690	760	750
Tear[e]	77	98	85	120	94
TD					
Modulus[c]	480	530	500	570	480
Tensile[d]	34	36	35	38	35
Elongation[a]	830	830	820	820	830
Tear[e]	280	220	190	180	230

Notes: a - units in %, b - dart drop impact in g, c - 1% secant modulus in MPa, d - tensile strength @ break in MPa, e - Elmendorf tear strength in g

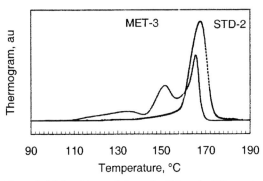

Figure 3. DSC melting endotherm for Z-N and miPP samples by Method B (SIST) heated at 20°C/min.

Introduction section, might be inherent to the presence of 2-1 insertion defect. It implies that the miPPs tend to crystallize the thinner lamellae and/or less perfect crystallites. The narrow half-width might come from the narrow distributions of the 2-1 insertion and racemic defects. It is noted that for phenomena of melting peak depression the MWD effect, i.e., chain ends as defects, can be ignored in the range of MW over 10,000 g/mol.[8]

The thermograms according to Method B (SIST) are presented in Figure 3 for the miPP and Z-N samples. The miPP is resolved to have three endotherms, whereas only a single melting peak for the Z-N iPP. The miPP might have possibly more peaks by conducting the isothermal crystallization in the wider temperature range. The three peaks formed represent

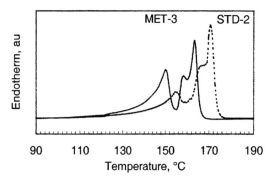

Figure 4. DSC first melting behavior for the BOPP films of Z-N and miPP resins heated at 20°C/min.

distinct families of homogeneous macromolecules in defect type, composition and distribution. Each family forms lamellar crystallites of different thickness and internal perfection.[6,7]

The single melting peak appearing at above 165°C for both the samples seems to be the monoclinic α crystal form, which is thermodynamically most stable and common in iPP.[9,10] The miPP appears to have less than a half of the a phase with the narrow distribution in thickness. The Z-N resin indicates only the α form iPP with the broad distribution in thickness. The complication of multiple endotherms is not observed, which often arises in iPPs from various effects, for the ZN resin because it has high MW and tacticity and because of the Method B conditions including high crystallization temperatures and long crystallization time.[11]

The lower melting endotherms, only showing in the miPP, can arise from: different polymorphic forms such as β, discrete morphological population of the α form but differing in size and perfection, or segregation effects by tacticity and defect distribution.[11] Nevertheless, the lower endotherms of the miPP are probably due to the unique presence of the 2-1 insertion defect. It is interesting that a distinct peak of about 150°C is at the proximity of the pseudo-hexagonal β crystal form, which is usually found in iPPs that have been subjected to high orientation or deformation in the melt.[9] This crystal structure will be investigated in the future. It is postulated that the miPP is effective for crystallization in the lower melting phase due to the presence of 2-1 insertion defect.

Figure 4 shows the first melting behavior of BOPP films scanned at 20°C/min for the miPP and Z-N samples. Multiple endotherms are observed for the both film samples. These endotherms, drawn at above 130°C to a greater degree of extension, may stem from the existence of distinct crystalline species with different levels of crystal disorder, crystal sizes, crystal forms or morphological forms.[12] In each film, the double peaks at higher temperatures are associated with the α crystal form with various degrees of crystal disorder and/or crystal sizes.[9] The miPP film seems to have the thinner lamellae with the less perfect crystallites than the Z-N iPP. The small crystalline size of the miPPs may be associated with the superior optical properties in quenched film applications to be discussed below. The higher melting points above that of the isotropic one (2nd melting) for each film imply to the higher degrees of crystal order and size due to the orientation. The lowest endotherm at about 155°C in the Z-N film

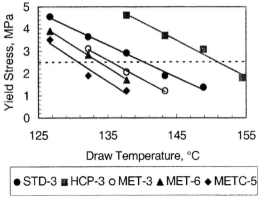

Figure 5. Effects of temperature on yield stress for Z-N and metallocene products.

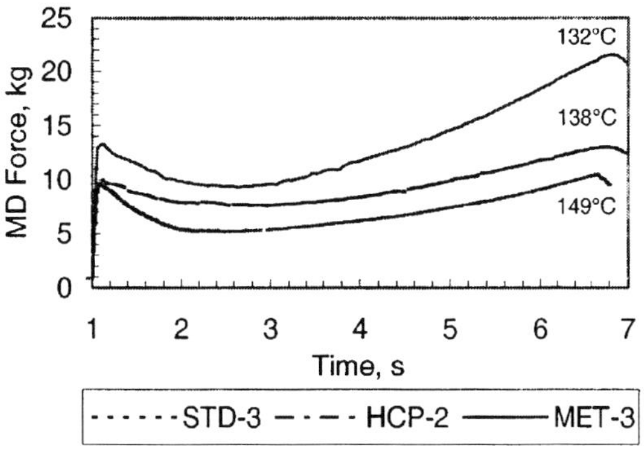

Figure 6. Biaxial stretching curves for samples selected from Figure 5 with comparable yield stress.

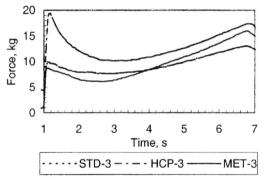

Figure 7. Biaxial stretching curves for samples selected from Figure 5 at draw temperature of 138°C.

sample may be assigned to the β form common in the highly deformed iPPs.[9] Although the highly oriented miPP has a significantly high fraction of the phase at 150°C, the β form is not detected using WAXS (wide angle x-ray scattering).

The BOPP processability can be correlated to the stress vs. strain behaviors obtainable using a laboratory scale film stretcher.[13,14] The effect of draw temperature on the yield stress is shown in Figure 5. The yield stress decreases linearly with increasing temperature. The gradual slopes and wider draw temperature range imply a good processing window. The yield stress is lower for all the miPPs at a given draw temperature. At an equivalent yield stress, the mJPPs require approximately 5-10°C lower draw temperature than the standard Z-N iPPs. This may be applied for a tenter frame process to control oven temperatures.

Figures 6 and 7 compare stress-strain curves for the three samples selected from Figure 5 at an equivalent yield stress and at the same draw temperature. Each sample deforms by a process of yielding, neck elongation and strain hardening. The miPPs yield rather rapidly compared to the gradual transformation of the standard ZN iPPs. For the yielding behavior, the miPPs are very similar to the high crystallinity iPP sample attributing to their high isotacticity. MET-3 noticeably exhibits an earlier onset and sharp increase in strain hardening. The earlier strain hardening and higher growth rate as well as the lower yield stress may relate to the relatively high fraction of the low melting species observed in the DSC thermograms. The high fraction of these low melting species of the miPP is largely crystalline in the MD orientation in the BOPP process, but it is much less crystalline at tenter oven conditions and may reduce the draw stresses required.[14] This high strain hardening effect may result in the enhanced film properties such as stiffness, COF and barrier discussed below.

Results of the BOPP and cast film properties are listed in Table 2 and 3, respectively. For the cast films, the miPPs show high MD tear strengths and superior optical properties to the Z-N iPP. The miPP homopolymers have a high stiffness. NMT-14 exhibits significantly improved overall properties. For the BOPP films, the miPPs show significant improvements in stiffness, COF, and MVTR (moisture vapor transmission rate), while satisfactory in optical properties. MET-3 exhibits a very low OTR (oxygen transmission rate). Optimizing process conditions would further enhance these film properties. The benefits of these properties have been described in the previous paper,[3] such as for a longer shelf-life of packaged foods, better down-stream handling, aesthetic appeal, down gauging, etc.

CONCLUSIONS

The DSC fractionation method gives possibilities to miPPs for identification of distinct families of molecular species. There are at least three families of species in the miPP examined here. It is postulated that the miPP is effective for crystallization in the lower melting phase due to the presence of 2-1 insertion defect, which is not present in the Z-N iPPs. The high fraction of the lower melting species might explain behaviors of BOPP processing and superior film properties demonstrated in the cast and BOPP film applications.

ACKNOWLEDGMENTS

We owe special thanks to Ben Hicks for his work of the cast film and TMLong® processes, and Rebecca Davis for the DSC measurements. We acknowledge Fina Oil and Chemical Company for the support and approval to publish this paper.

REFERENCES

1   **Metallocene Polypropylene 1995**, *Phillip Townsend Associates Inc.*, Houston, TX, 1995.
2   K. Hungenberg, J. Kerth, F. Langhauser, H. Muller, and P. Muller, *Makromol. Chemie*, **227**, 159 (1997).
3   R. Wheat and A. Hanyu, SPO '97 (1997).
4   K. Hungenberg, J. Kerth, F. Langhauser, B. Marczinger, and R. Schlund, **Ziegler Catalysts**, G. Fink, R. Mulhaupt, and H. H. Brintzinger, ed., *Springer Verlag*, Heidelberg, 363 (1995).
5   H. J. Zimmermann, *J. Macromol. Sci.*, **B32**, 141 (1993).
6   E. Adisson, M. Ribeiro, A. Deffieux, and M. Fontanille, *Polymer*, **33**, 4337 (1992).
7   P. Starck, *Polymer International*, **40**, 111 (1996).
8   T. Kamiya, N. Ishikawa, S. Karnbe, N. Ikegami, H. Nishibu, and T. Hattori, ANTEC '90, 871 (1990).

9   S. Vleeshouwers, *Polymer*, **38**, 3213 (1997).
10   A. J. Ryan, J. L. Stanford, W. Bras, T. M. W. Nye, *Polymer*, **38**, 759 (1997).
11   R. A. Phillips and M. D. Wolkowicz, **Polymer Handbook**, E. P. Moore, Jr., ed., Hanser, Munich, 1996.
12   R. J. Yan and B. Jiang, *J. Polym. Sci. Polym. Phys.*, **31**, 1089 (1993).
13   E. K. Bullock and W. W. Cox, TAPPI '95, 571 (1995).
14   W. W. Cox and E. K. Bullock, Polyolefins RETEC'97, 397 (1997).

Phase Structure Characterization and Processing-Structure-Property Relationships in Linear Low-Density Polyethylene Blown Films

Jianjun Lu, Baiyi Zhao and Hung-Jue Sue

Polymer Technology Center, Department of Mechanical Engineering, Texas A&M University, College Station, TX 77843-3123

INTRODUCTION

Processing-structure-property (P-S-P) relationships in polyethylene (PE) blown films have been studied for more than four decades. The purpose of such study is to enable the manufacturers to predict the processing conditions required for achieving certain film properties for a certain PE resin. Although there is a wealth of knowledge generated by the polyolefins industry related to the production of PE blown films, this wealth of knowledge is largely based on empirical and semiempirical relationships established for known systems. When the polyolefin structure is altered, the pre-existing empirical or semi-empirical knowledge is no longer valid. The fundamental understanding concerning how to relate the molecular structure and processing condition to the morphology and properties of blown films is still lacking.

In film blowing process, the primary molecular parameters, such as molecular architecture, molecular weight and its distribution, and processing condition can all affect film morphology. The morphological structure of film, in turn, determines the final film properties. Therefore, the characterization of film morphological structure is crucial for the understanding of P-S-P relationships.

It is known that a semi-crystalline polymer consists of crystalline, crystalline-amorphous interfacial and amorphous regions. The tubular film blowing process is characterized by a very complex structure formation caused by different parallel crystallization and orientation processes.[1] The resultant film morphology is very different from that of bulk material. Much research has been done on the morphological characterization of PE blown films. In 1954, Keller[2,3] proposed that the crystal b-axis in PE blown films has a preferred orientation perpendicular to the machine direction (MD) while the a- and c-axes are randomly distributed with cylindrical symmetry. He also raised a concept of row orientation in which the crystalline lamellar overgrowth occurs epitaxially from the c-axis oriented fibrils with radial growth in the b direction and the a- and c-axes rotating about that growth direction. Using the method of pole figures for HDPE extruded films, Lindenmeyer and Lustig found support for the row structure.[4] In 1967 Keller and Machin[5] further modified the model of "row nuclei" based on the level of stress for both LDPE and HDPE extruded films. According to the modified model, two major crystallizations take place depending upon the magnitude of the stress in the melt, namely "low-stress" and "highstress" crystallizations. They suggested that the row nucleated structure is similar to the "shish-kebabs" crystallized from the stirred solution state.

In the 1970's, Maddams and Preedy published a series of papers based on a very extensive texture investigation on very different HDPE blown films.[6-8] Noticeably, they found an additional orientation maximum of b-axis in the normal direction (ND) connected with a transcrystalline portion of material. The finding of the transcrystallization process was confirmed by Eby,[9] Fitchmun and Newman.[10]

Choi, Spruiell and White[11] used WAXS, SAXS and birefringence to characterize films fabricated under conditions ranging from uniaxial to biaxial extension, with concurrent measurement of the machine direction and transverse direction (TD) stresses at the frost line. They proposed morphologies consisting of local lamellar stacks with a distribution of stack orientations that is determined by the MD/TD stress balance at crystallization.

Although the crystalline morphology characterization led to a great improvement in the understanding of the morphological structure of PE blown films, knowledge about the crystalline phase is not sufficient for making a structure-property correlation. The effect of the structure of noncrystalline phase on film properties is also very significant. Although the noncrystalline phase is composed of a noncrystalline interfacial region with limited molecular mobility and a rubbery amorphous region with high molecular mobility, the PE blown films in most previous studies were just treated as composites of crystalline lamellae and amorphous material. There are a few publications concerning amorphous phase orientation in PE blown films.[12-14] Some researchers utilized solid-state NMR[15,16] and Raman[17,18] spectroscopy to study noncrystalline interfacial region in bulk crystallized and/or solution crystallized polyethylene samples. However, not enough attention has been given to the characterization of crystalline-amorphous interphase in the films.

The previous investigations on polyethylene blown films were mainly carried out on high-density polyethylene (HDPE), due to its simplicity of molecular architecture. Nevertheless, the blown films based on LLDPE give improved transparency and mechanical properties. Moreover, recent developments of new catalysts (metallocene catalysts) and processes enable producers to exercise greater control over short chain branching distribution and molecular weight distribution. At this stage, no extensive morphological investigation has been performed on LLDPE blown films as on HDPE ones.

As part of a larger effort to investigate P-S-P interactions in polyethylene blown films, this study chose two LLDPE blown films made under different processing conditions. Morphology investigation was carried out using solid-state NMR, DSC, TEM, and IR dichroism. Morphological features are then correlated with the mechanical properties and processing condition.

EXPERIMENTAL

MATERIALS

Two LLDPE blown films (Films A and B) were provided by Phillips Petroleum Company. The resin has a density of 0.922 g/cm^3 and a melt index of 0.22 g/10 min. The thickness of the films is around 25 μm. Film A was made by using a conventional bubble geometry for LLDPE (i.e., a low stalk bubble configuration), while Film B was blown by using a high stalk bubble configuration.

SOLID-STATE NUCLEAR MAGNETIC RESONANCE SPECTROSCOPY

^1H diplor-decoupling/magic-angle spinning ^{13}C NMR measurements were carried out at room temperature on a MSL-300 spectrometer operating at a field strength of 7.0 T. Each sample was packed into a 7 mm diameter zirconia rotor. The magic angle spinning was performed at 4.4 to 4.6 kHz. ^{13}C NMR chemical shifts were expressed as values relative to tetramethylsilane (Me_4Si). ^{13}C NMR spin-lattice relaxation time (T_1) was measured by inversion recovery pulse sequence with cross-polarization. The recycle delay for T_1 measurement was 2500 s and the number of scans per spectrum was 8. ^{13}C NMR spin-spin relaxation time (T_2) measurement was performed by spin-echo pulse sequence. The recycle delay for T_2 experiment was 2 s.

DIFFERENTIAL SCANNING CALORIMETRY

The DSC experiments were performed on a Perkin Elmer Pyris-1 differential scanning calorimeter. Each experiment includes 3 steps: 1) temperature scan from -30°C to 180°C at a heating rate of 10°C/min; 2) cooling from 180°C to -30°C at a cooling rate of 20°C/min; and 3) repeating the first step. A sample weight ranging from 7 to 10 mg was used.

TRANSMISSION ELECTRON MICROSCOPY

Samples for TEM study were embedded in an epoxy and cured overnight at room temperature. The blocks were trimmed, faced-off, and then stained with RuO_4. Ultra-thin sections, ranging from 60 to 80 nm, were obtained using a Reichert-Jung Ultracut E microtome with a diamond knife. The thin sections were placed on 100-mesh Formvar-coated copper grid and examined using Zeiss-10C TEM operating at an accelerating voltage of 100 kV.

INFRA-RED DICHROISM

A Nicolet Avatar 360 FTIR equipped with a polarizer was used to determine IR dichroism. The absorption bands at 730 and 719 cm^{-1} were employed to evaluate the orientations of crystal a-axis and b-axis, respectively. The spectral separation procedure described by Kissin[19] was utilized to obtain the real absorbances from the measured absorbances of the bands. A dichroic ratio (D) was taken as the ratio of the absorbance measured with radiation polarized in MD to that with radiation polarized in TD. The method developed by Read and Stein[20] was used to approximate the orientation functions of crystal axes a and b, f_a and f_b, respectively, from the dichroic ratios. Once f_a and f_b are obtained, f_c can be calculated since the sum of f_a, f_b, and f_c must be zero.

RESULTS

CRYSTALLINE AND NONCRYSTALLINE CONTENTS

The weight percentage crystallinity of the PE films was obtained from DSC by comparing the heat of fusion with 100% crystalline polyethylene heat of fusion of 293 J/g.[21] From the solid-state ^{13}C NMR spectra, the crystalline, interfacial and rubbery phase contents in the blown films can be resolved by using the method described by Horii *et al.*[15] The crystalline and noncrystalline content data obtained on DSC and solid state NMR are summarized in Table 1.

LAMELLAR THICKNESS DISTRIBUTION

In principle, DSC melting endotherms can be used to calculate the lamellar thickness distribution. In this study, a DSC profile (heat flow versus temperature) is converted into a lamellar thickness distribution curve (relative probability of the weight percentage of lamellae versus lamellar thickness) by using the Gibbs-Thomson equation.[22] The lamellar thickness distribution curves are shown in Figure 1.

For each film, several independent DSC tests were performed. It is found that the shape of melting peak upon the first temperature scan, thus the lamellar thickness distribution in the film, varies significantly from test to test, although the heat of fusion stays constant. After the specimen was cooled to -30°C again, the second temperature scan was performed. The second scans, give melting peaks with the same shape in all tests on both films. This is due to that the thermal history in the original films was eliminated upon the first melting. This also indicates that the variation of melting peak shape upon the first temperature scan is not an experimental error, and the morphological structure is not uniform in the films. However, it is interesting to

Table 1. Morphological parameters and mechanical properties of films A and B

	Film A	Film B
Crystalline content, DSC	46	46
Crystalline content, NMR	59	52
Interfacial content, NMR	24	32
Amorphous content, NMR	17	16
D_{730}	1.852	1.405
D_{710}	0.317	0.580
f_a	0.221	0.119
f_b	-0.294	-0.162
f_c	0.073	0.043
MD tensile modulus, MPa	144.4 @ 1%	102.8 @ 1%
TD tensile modulus, MPa	226.2 @ 1%	102.8 @ 1%
MD Elmendorf tear, g	14	81
TD Elmendorf tear, g	666	480
Dart impact resistance, g	53	341

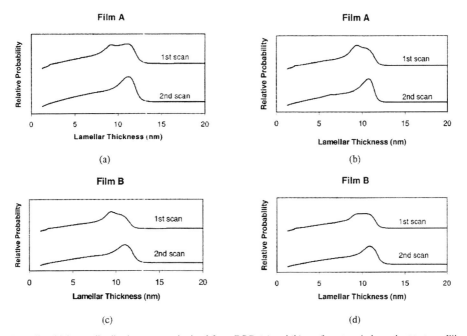

Figure 1. Lamellar thickness distribution curves obtained from DSC: (a) and (b) are from two independent tests on Film A; (c) and (d) are from two independent tests on Film B.

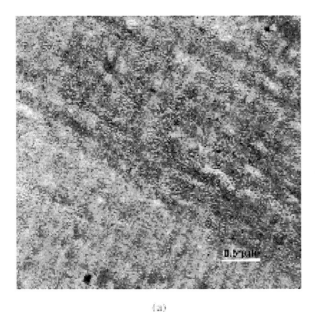

(a)

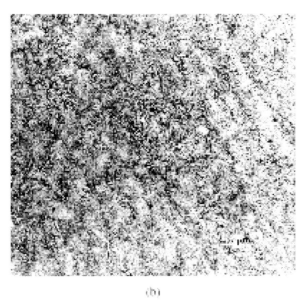

(b)

Figure 2. TEM micrographs of (a) Film A and (b) Film B.

note that each of these melting peaks is seemingly composed of two overlapped peaks with different peak temperatures. The magnitudes of these two peaks vary, but the sum of the two is relatively constant. Since each melting temperature corresponds to a lamellar thickness, it appears that there exists a bimodal lamellar thickness distribution.

CRYSTALLINE ORIENTATION

Figures 2a-2b are TEM micrographs taken in the MD-ND plane of Films A and B. It is clearly shown that Film A possesses a significant higher lamellar orientation than Film B. Film B exhibits a rather random crystalline texture, but a lamellar orientation with the lamellar normal preferentially aligned with MD can still be observed.

The IR dichroic ratios and calculated orientation functions are listed in Table 1. IR dichroism results suggest that the crystallites in Film A are more orientated than in Film B. This result is consistent with the TEM result.

DISCUSSION

High stalk bubble configuration is somehow novel for LLDPE film blowing as it is normally used for HDPE blown films. The region between the die face and the stalk (bubble expansion zone) provides a region of low elongation rate in which a portion of the MD orientation introduced in the die can be relaxed.[23-25] A higher stalk height gives a longer time for orientation relaxation, thus less MD orientation is "frozen" into the final blown films. This is the reason why Film A possesses a higher chain backbone orientation (c-axis orientation, f_c) than Film B.

The use of the high stalk bubble configuration has another effect on the crystallization and orientation processes. For this bubble configuration, the film blowing process is divided into two distinct deformation regimes,[25] namely, a nearly planar extension in the stalk and a biaxial stretching in the bubble expansion zone. In the stalk region, a significant portion of MD stretching takes place, and the bubble diameter decreases (neck-in).[26] Consequently, the magnitude of TD strain from the neck to the frost line can be much larger than the overall blow-up ratio, and the overall MD/TD stretch ratio becomes closer to 1. So, Film B has a more random lamellar orientation compared with Film A.

For convenience, some film mechanical properties obtained from the manufacture are also listed in Table 1. From the table, we can see that tear resistance can be correlated with the chain orientation. The crystal c-axes in both films are preferentially oriented in MD. This results in a larger Elmendorf tear strength in TD for both films. As the c-axis orientation in Film B decreases, the difference between the tear strengths in MD and TD decreases. This may be why Film B has a higher dart impact resistance than Film A, since the dart drop impact resistance is believed to be related to the balance of mechanical properties between MD and TD.

The effect of crystalline orientation on Young's modulus can be well illustrated by the "laminated composite" model.[27] Considering a film in which crystalline lamellae are uniaxially oriented with lamellar normal preferentially aligned parallel to MD, it can be viewed as a "laminated composite" of a hard crystalline phase and a soft amorphous phase stacked along MD. If the film is stretched in MD, the material can be considered as the composite being deformed in an isostress pattern; therefore, the modulus is dominated by that of the soft amorphous phase. On the other hand, if the film is stretched in TD, the material can be considered as the composite being deformed in an isostrain pattern; therefore, the modulus is dominated by that of the hard crystalline phase. However, the modulus difference between the two films in this work can hardly be explained in terms of crystalline orientation, because film A has a higher modulus than Film B in both MD and TD. This leads us to consider the effect of crystallinity.

Surprisingly, the crystalline contents measured by DSC are same for the two films. Similar results were obtained on some HDPE blown films made from the same resin, but which differ by bubble stalk height.[28] However, the solid-state NMR results show a significant difference between the two films in terms of crystalline and interfacial contents. Film A has a higher crystalline content and a lower interfacial content than Film B. The rubbery contents in these two films are similar. The higher modulus of Film A seemingly can be attributed to its higher crystalline content.

It is worthwhile to mention that DSC and solid-state NMR rely on different principles to measure the crystalline and noncrystalline contents. While DSC measures the energy required to destroy the order in crystalline phase, NMR distinguishes crystalline, interfacial and amorphous regions by the difference in molecular mobility. The disagreement arises from this difference in measurement mechanism. We are planning to investigate this issue further.

There are several possible reasons for the bimodal distribution of lamellar thickness. First, the two most probable lamellar thickness peaks might correspond to crystallites formed in two different regions of the bubble during the film blowing process, i.e., stalk region and bubble expansion region. Second, different parallel crystallization processes may take place in film blowing. While the stress crystallization is caused by nucleation in the volume, for the blown films there also exists an additional possible crystallization mechanism, i.e., the transcrystallization process,[1] in which crystallization starts from the nuclei at the surface of the melt. This results in a preferential orientation of b-axis towards the ND of the film. These two different crystallization processes may produce crystallites with different sizes. Another possible reason is the temperature gradient across the thickness of the film. The crystallization temperature may be different between the interior and the surface, or between the inside and outside surfaces of the bubble. It will be of interest to figure out the nature of this bimodal lamellar thickness distribution.

CONCLUSIONS

Compared with the conventional bubble geometry for LLDPE blown films, the high stalk bubble configuration can result in a more random lamellar texture and more balanced mechanical properties. DSC and solid-state NMR gave significantly different crystalline contents for the same LLDPE film in this study. The results of NMR can be used to interpret the modulus difference between the two films, which otherwise can not be explained by DSC or by the crystalline orientation data alone. A bimodal lamellar thickness distribution is found for both films studied. Investigating this bimodal lamellar thickness distribution would give more insight to the crystallization mechanisms during the film blowing process.

ACKNOWLEDGMENTS

The authors would like to thank Polyolefins Films Consortium for financial support and providing PE blown films. Special thanks are also given to Mr. E. I. Garcia-Meitin for his valuable instruction on TEM work.

REFERENCES

1   D. Hofmann et al., *J. Appl. Poly. Sci.*, **39**, 1595 (1990).
2   A. Keller, *Nature*, **174**, 826 (1954).
3   A. Keller, *J. Polymer Sci.*, **15**, 31 (1955).
4   P. H. Lindenmeyer and S. J. Lustig, *J. Appl. Polym. Sci.*, **9**, 227 (1965).
5   A. Keller and M. J. Machin, *J. Macromol. Sci.*, **B1**, 41(1967).
6   W. Maddams and J. Preedy, *J. Appl. Polym. Sci.*, **22**, 2721 (1978).
7   W. Maddams and J. Preedy, *J. Appl. Polym. Sci.*, **22**, 2738 (1978).
8   W. Maddams and J. Preedy, *J. Appl. Polym. Sci.*, **22**, 2751 (1978).
9   R. K. Eby, *J. Appl. Phys.*, **35**, 2720 (1964).
10   D. Fitchmun and S. Newman, *J. Polym. Sci.*, **B7**, 301(1969).
11   K. Choi, J. Spruiell, and J. White, *J. Polym. Sci., Polym. Phys. Ed.*, **20**, 27 (1982).
12   R. M. Patel et al., *Polym. Eng. Sci.*, **34**, 1506 (1994).
13   A. M. Sukhadia, SPE ANTEC, 160 (1998).
14   Y. Kim and J. Park, *J. Appl. Polym. Sci.*, **61**, 2315 (1996).
15   R. Kitamaru, F. Horii, and K. Murayama, *Macromolecules*, **19**, 636 (1986).
16   K. Kuwabara, et al., *Macromolecules*, **30**, 636 (1997).
17   G. R. Strobl and W. Hagedorn, *J. Polym. Sci., Polym. Phys. Ed.*, **16**,1181 (1978).
18   M. Glotin and L. Mandelkern, *Colloid & Polymer Sci.*, **260**, 182 (1982).
19   Y. V. Kissin, *J. Polym. Sci., Polym. Phys. Ed.*, **30**, 1165 (1992).
20   B. E. Read and R. S. Stein, *Macromolecules*, **1**(2), 116(1968).
21   L. Mandelkern, **Crystallization of Polymers**, *McGraw Hill*, New York (1964).
22   J. D. Hoffman, **Treatise on Solid State Chemistry**, Vol. 3, *Plenum Press*, New York (1976).
23   H. H. Winter, Pure and Appl. Chem., 55, 943 (1984).
24   T. Kanai, M. Kimura, and Y. Asano, *J. Plastic Film and Sheeting*, **2**, 224 (1986).
25   P. P. Shirodkar and S. D. Schregenberger, SPE ANTEC, 37 (1987).
26   E. J. Dormier et al., SPE ANTEC, 696 (1989).
27   H.Zhou and G. L. Wilkes, *J. Mater. Sci.*, **33**, 287 (1998).
28   Unpublished data.

The Relative Influences of Process and Resin Time-Scales on the MD Tear Strength of Polyethylene Blown Films

Rajendra K. Krishnaswamy and Ashish M. Sukhadia
Phillips Petroleum Company, 94-G, Bartlesville, OK 74004, USA

BACKGROUND

Commercial production of polyethylene in the United States alone is more than ten million metric tons, of which approximately 30% is LDPE, 25% is LLDPE, and 45% is HDPE. Besides, LLDPE is at the onset of a major growth process initiated by the advent of single site catalysts that allow substantially better control of the polymer's molecular architecture. Almost 70% of the commercial LLDPE and LDPE are used in the form of blown and/or cast films.[1]

It is recognized that the orientation state of the crystalline and non-crystalline phases largely dictates the performance of blown films in their intended application. We are aware of the fact that the ultimate morphology of a blown film develops over a short region in the vicinity of the process frost line. Also, the final semi-crystalline morphology that develops is very likely to be dependent on the state of the macromolecules prior to the onset of crystallization. The state of the molecules prior to and during the crystallization process, in turn, is dependent on the influence exerted by the process stress profiles in orienting the macromolecules and subsequently either preserving or relaxing their extended state. Thus, if the ultimate performance of blown films is a function of their orientation characteristics, what happens to the macromolecules as they exit the die and subsequently crystallize under extensional deformation will obviously have an impact on the performance of the ultimate product. Consequently, our objective here is to identify possible morphological features that drive the performance of polyethylene blown films, specifically, the MD tear strength.

BLOWN FILM PROCESS MD EXTENSION RATE

Momentum balance of the blown film process leads to an equation for the extension rate profile along the bubble as a function of distance from the die exit. The extension rate along the film MD is expressed by Ghaneh-Fard *et al.*[2,3] as follows:

$$\dot{\varepsilon}_{MD} = 2\frac{dv_z}{dz} + 2v_z\left(\frac{dr}{dz}\right)\left(\frac{d\theta}{dz}\right) \qquad\qquad [1]$$

where v_z is the axial film velocity, r is the bubble radius, and θ is the bubble inflation angle. Video tracing techniques are typically required to measure strain rate profiles along the bubble. However, as the video tracing method is difficult to incorporate with every film run, the machine direction extension rate at the frost line can be approximated as follows (from equation 1):

$$\dot{\varepsilon}_{MD} \approx 2\frac{v_{FL} - v_{die}}{FLH} \qquad\qquad [2]$$

where the subscripts FL and die indicate frost-line and die exit, respectively, and FLH is the frost-line height. The above equation is a useful tool to point differences or magnitude of differences in the process MD extension rates of films blown under different conditions.

EXPERIMENTAL

Firstly, a single lot of Phillips low density linear polyethylene (LDLPE; pellet density = 0.924 g/cc) resin was blown into film at various processing conditions. It is important to note that this LDLPE resin is different from conventional LLDPE resins in that it is very broad in molecular weight distribution ($M_w/M_n \sim 20$). All the films were blown at high-stalk conditions and the thickness of all films was approximately 25 microns (1 mil). The process conditions that were varied were blow-up ratio, extrusion rate and temperature, frost-line height and die gap. This provided us with several blown films (from a single resin) varying in process strain rate profiles and ultimate performance. This sample set will be referred to as SERIES-1.

In a second series, several experimental LDLPEs were blown in the in-pocket (conventional LLDPE) configuration. These LDLPEs were of comparable density, molecular weight, molecular weight distribution, and melt index. However, the melt relaxation times of these resins varied between 3 s and 25 s. The spread in relaxation times is speculated to be due to the presence of different levels of long chain branching in the polymer. The relaxation times of these resins are based on Carreau-Yasuda fits to small-amplitude oscillatory measurements at 190°C; the details of the data treatment can be found elsewhere.[4] Each of these resins was converted into two films of different thickness, with all other processing conditions remaining constant. This series will be referred to as SERIES-2. SERIES-2 contains two subsets; one processed at high extension rates (75 micron thickness) and the other processed at low extension rates (125 micron thickness), with all other processing conditions remaining constant.

All of the film blowing was carried out using a 38 mm diameter single screw Davis Standard extruder (L/D = 24; 2.2:1 compression ratio), which is fitted with a barrier screw with a Maddock mixing section at the end. A 5.1 cm diameter Sano film die equipped with a single lip air ring was employed for SERIES-1, and a 10.2 cm diameter Sano film die equipped with a dual lip air ring was employed for SERIES-2. For SERIES-2, the temperature setting along the extruder and die were 220°C, the blow-up ratio was 15, the die gap was 60 mm, and the extrusion rate was approximately 27 kg/hr.

For all blown films, the approximate MD extension rate at the frost line was estimated as shown in the previous section (equation 2). For SERIES-1, the White/Spruiell biaxial orientation factors of the crystalline phase were estimated by a method described elsewhere.[5] The biaxial orientation factors of the non-crystalline phase for some select SERIES-1 blown films were estimated by coupling the crystalline orientation data with birefringence (in-plane and out-of-plane) measurements as documented in literature.[6-9] For blown film birefringence data, refractive index measurements were obtained using a Metricon Prism Coupler instrument. The MD tear strengths of all blown films were measured according to ASTM D1922.

RESULTS AND DISCUSSION

In Figure 1, the MD tear strength of the SERIES-1 blown films are plotted as a function of the average MD extension rate at the frost-line as estimated using equation 2. Here, the films with the highest MD tear strengths were processed at low MD extension rates. At low extension rates, the stresses experienced by the polymer will be lower relative to processing at higher extension rates. Thus, it appears that high extensional

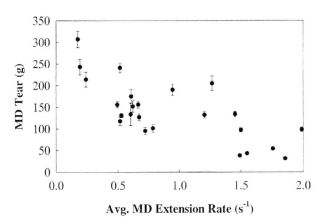

Figure 1. MD tear vs. average MD extension rate.

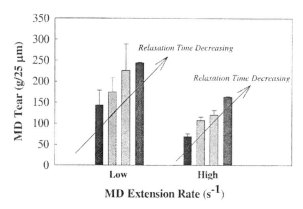

Figure 2. Series-2. Tear vs. extension rate.

stresses along the MD of a blown film process are detrimental from a MD tear performance perspective. At high MD extension rates, the degree of molecular orientation prior to crystallization is anticipated to be high. This can result in preferential orientation of the chain segments along the MD and this imbalance in orientation can have significant implication on the tear performance of the corresponding blown film.

The extension rate dependence discussed above suggests that the blown film MD tear strength of LDLPE resins, at fixed process conditions, is a function of the resin melt viscosity and/or relaxation time as measured under planar or biaxial extensional deformation. In order to test this hypothesis, several LDLPE resins of different melt relaxation times and comparable molecular weights (and distribution) were blown into film at two separate conditions and their properties were compared (SERIES-2, Figure 2). It should be cautioned that the relaxation times of the resins in SERIES-2 were measured under dynamic shear deformation, while similar estimates under extensional deformation are more representative and hence preferred. With this in mind, we clearly observe (Figure 2) that the MD tear strength of the blown film increases systematically with decreasing relaxation times. This trend is evident for both sets of films which differ only in the draw-down ratios (ultimate thickness) and hence average MD extension rate at the frost line. In Figure 2, we also observe that the films processed at lower extension rates consistently display higher MD tear strengths for all four LDLPE resins. Thus, it is clear that not only is the relaxation time of the resin important, but its relative magnitude with respect to the timescale of the process is also critical. In other words, when the Deborah number (relaxation time/process time) is low, the MD tear strength of the resulting blown films will be high.

In LDLPE resins, the melt relaxation times can increase due to the presence of long branches or a fraction of substantially high molecular weight molecules. Thus, we conclude that both long, branches and small amounts of very high molecular weight molecules are both detrimental from a blown film MD tear performance perspective.

The Elmendorf tear test is a process of mechanical deformation at relatively high rates. Such a deformation process involves substantial molecular rearrangement in the crystalline and non-crystalline phases. Some of the more prominent deformation mechanisms identified

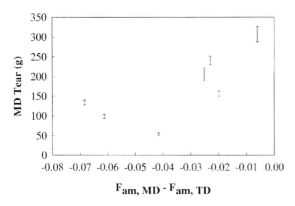

Figure 3. Effect of orientation factors on tear.

during the tensile deformation of polyethylene are interlamellar shear, interlamellar separation, intralamellar chain slip and lamellar rotation.[10-17] These local deformation processes are usually followed by break-up of the lamellar structures leading to catastrophic failure of the material. During the initial stages of deformation, the non-crystalline phase ("softer" phase) deforms more readily relative to the crystalline phase. Thus, during the initial stages, interlamellar shear and interlamellar separation are probably dominant. At high strain levels, the crystalline phase begins to deform presumably through chain slip and lamellar rotation.

In a tear test, some or all of the above-mentioned deformation mechanisms can contribute in providing resistance to tear propagation. In Figure 3, the MD tear strength of some of the SERIES-1 blown films are plotted as a function of the difference between the MD and TD orientation factors of the non-crystalline chain segments. The greater the magnitude of this difference ($F_{am, MD} - F_{am, TD}$), the greater is the tendency for the non-crystalline chains to be located preferentially along either the MD or the TD. On the other hand, a value of ($F_{am, MD} - F_{am, TD}$) closer to zero indicates equi-biaxial orientation of the non-crystalline chains with respect to the film MD and TD. In Figure 3, we observe that the films with the highest MD tear strengths are those in which ($F_{am, MD} - F_{am, TD}$) is closest to zero. In other words, the closer the noncrystalline chain segments are to equi-biaxial orientation, the higher is the resistance to tear propagation along the film MD. Similar trends (not reported here) were observed for various LLDPE resins blown into film at similar conditions.

Interlamellar shear was shown to be the dominant initial mode of deformation when the tensile stretching direction was 45° with respect to the chain axis in the crystalline phase of HDPE films with well defined stacked lamellar textures.[15,16] In the blown films considered here, the resistance to MD tear propagation appears to be higher when the non-crystalline chain segments are aligned at approximately equal angles with respect to the MD and TD. This suggests that interlamellar shear plays an important role in determining the MD tear performance of LDLPE blown films. Interlamellar shear deformation has also been associated with the double yielding characteristics observed in some polyethylene films.[18,19] LDLPE blown films, in general, reveal a distinct double-yielding behavior when deformed in the tensile mode along the MD. Therefore, it is reasonable to anticipate that interlamellar shear can play a noticeable role when these blown films are deformed along their MD. However, further research is required to determine if interlamellar shear is the limiting factor that determines the MD tear performance of LDLPE and LLDPE blown films. It is cautioned that other molecular factors and orientation features, not considered here, may influence the MD tear performance of polyethylene blown films as well.

Finally, we know that the polymer molecules close to the die exit (neck region) tend to be highly oriented along the MD. In order for the non-crystalline chain segments to be oriented equi-biaxially with respect to the film MD and TD, they will not only need to relax their extended conformation to a certain extent but they will also be required to respond to the transverse stresses exerted during the blow-up. Therefore, in situations wherein the MD extension rate is very high relative to the TD extension rate or when the relaxation time of the resin is substantially longer than the process time-scale, the probability of attaining equi-biaxial orientation of the non-crystalline chain segments along the plane of the film is low. Consequently, we observe that the dependence of LDLPE blown film MD tear on the process and resin time scales is reasonably consistent with the proposed interlamellar shear deformation mechanism.

CONCLUSIONS

For several blown films made from a broad molecular weight distribution, linear low density polyethylene type material (LDLPE), it was clearly shown that the MD tear performance was favored by low process MD extension rates and short polymer melt relaxation times. The relative magnitude of the polymer relaxation time and process time was also shown to be important, with low Deborah numbers favoring better MD tear performance. Orientation of the noncrystalline chain segments in these blown films appears to suggest that interlamellar shear plays an important role in determining the MD tear performance of LDLPE and LLDPE blown films.

ACKNOWLEDGMENTS

The encouragement and assistance offered by Prof. Garth Wilkes (VPI&SU) is greatly appreciated. David Rohlfing and Mike Hicks are appreciated for the rheological characterization of the resins. Jay Janzen and David Register are acknowledged for all their encouragement and useful discussions. Jerry Stark, David Higbee and John Wehmeyer were responsible for all the film blowing and testing. John Jackson of Metricon Corporation is acknowledged for refractive index measurements on the LDLPE blown films. The authors also express their gratitude to the Phillips Petroleum Company for encouragement and for permission to publish and present this work.

REFERENCES

1   J. F. Ross and J. L. MacAdams, "Polyethylene (Commercial)", from **The Polymeric Materials Encyclopedia**, pg. 5953-5965, *CRC Press* (1996).
2   A. Ghaneh-Fard, P. J. Carreau, and P. G. Lafleur, *Polymer Engineering and Science*, **37**, 1148 (1997).
3   A. Ghaneh-Fard, P. J. Carreau, and P. G. Lafleur, *International Polymer Processing*, **12**, 136 (1997).
4   J. Janzen, D. C. Rohlfing, and M. J. Hicks, *Journal of Rheology*, In Press (1998).
5   R. K. Krishnaswarny, SPE-ANTEC Proceedings, (1999).
6   J. L. White and J. E. Spruiell, *Polymer Engineering and Science*, **21**, 859 (198 1)
7   J. L. White and M. Cakmak, *Advances in Polymer Technology*, **8**, 27 (1988).
8   G. L. Wilkes, "Rheooptical Properties", from **Encyclopedia of Polymer Science and Engineering**, Mark, Bikales, Overberger, & Menges editors, vol 14, 261-298 (1985).
9   R. J. Pazur and R. E. Prud'homme, *Macromolecules*, **29**, 119 (1996).
10 A. Keller and D. P. Pope, *Journal of Material Science*, **6**, 453 (1971).
11 D. P. Pope and A. Keller, *Journal of Polymer Science: Polymer Physics Edition*, **13**, 533 (1975).
12 A. Peterlin, *Colloid and Polymer Science*, **265**, 357 (1987).
13 L. Lin and A. S. Argon, *Journal of Material Science*, **29**, 294 (1994).
14 R. S. Porter and L. H. Wang, *Journal of Macromolecular Science - Macromol. Chem. Phys.*, **C35**, 63 (1995).
15 H. Zhou and G. L. Wilkes, *Macromolecules*, **30**, 2412 (1997).
16 H. Zhou and G. L. Wilkes, *Journal of Material Science*, **33**, 287 (1998).
17 M. F. Butler and A. M. Donald, *Macromolecules*, **31**, 6234 (1998).
18 N. W. Brooks, R. A. Duckett and I. M. Ward, *Polymer*, **22**, 1975 (1992).
19 N. W. Brooks, A. P. Unwin, R. A. Duckett and I. M. Ward, *Journal of Macromolecular Science: Physics Edition*, **34**, 29 (1995).

Metallocene Catalyzed Polyethylene in Blown Film Applications. A Comparison Between Monoextruded Blended Films and Coextruded Films

C. M. Beagan, G. M. Mc Nally & W. R. Murphy

Polymer Processing Research Center Department of Chemical Engineering, The Queens University of Belfast, David Keir Building, Stranmillis Road, Belfast BT9 5AG, N. Ireland.

INTRODUCTION

Metallocene catalysts are finding ever increasing application due to their ability to create polymers with properties that can be precisely controlled over a wide range of limits.[2] Metallocene catalyzed polyethylenes are reported to have superior properties over conventional polyethylenes, principally due to their narrow molecular weight distribution and more uniform co-monomer distribution.[3] However, these polymers are more difficult to process because of the higher viscosity's encountered during melt processing.[4] Films produced from these resins are also susceptible to blocking.[5] However, in order to overcome some of these difficulties, mPE's may be blended or coextruded with other conventional polyethylenes.

The blending of polymers to achieve a superior product is used widely in industry. Many studies have been conducted on blends of LDPE and linear low density polyethylene (Linear Low Density Polyethylene).[6-10] These have shown that addition of increasing amounts of LLDPE will generally result in an increase in tensile properties. Martinez et al.[11] have compared blends of LDPE/LLDPE with coextruded films containing LLDPE as the middle layer and LDPE as the inner and outer layers and found, that in general the coextruded films showed the superior mechanical properties.

The aim of this paper is to determine the best way to utilize these resins in the blown film process to achieve optimum mechanical properties. Monolayer blended films of mPE and LDPE were compared to three layer coextruded films of A:B:A structure where A is LDPE and B is mPE. Crystallinity of the blended films was studied and the rheological behavior was also considered.

EXPERIMENTAL PROCEDURES

MATERIALS SELECTION

For this study two types of polyethylene, suitable for thin film gauges were used, a low density polyethylene (LDPE) from Exxon (MFI 2 g/l0 min) and a mPE grade from BASF (MFI 1.4 g/l0 min). For the blended materials a Shulmann antiblocking agent was added.

SAMPLE PREPARATION

The films were manufactured using a Killion blown film coextrusion line comprising one 38 mm extruder and two 25 mm extruders, fitted with a 75 mm diameter die, die gap of 2.0 mm. Barrel temperature profiles were maintained throughout the trials from 160°C at the feed section to 200°C at the die. Films were produced at a constant blow up ratio of 2.0 and a constant screw speed of 27 rpm set to produce a film thickness of 50 μm at a haul off speed of 3 m/min. To reduce film thickness haul off speed was then increased to 6 m/min, 9 m/min and l2 m/min and samples taken at each stage. At the highest haul off speed film thickness was approximately 12 μm.

Blends of LDPE and mPE were tumble mixed prior to extrusion in the following w/w ratios 80/20, 60/40, 40/60, 20/80. Previous experience at these laboratories has shown that films manufactured from mPE's are susceptible to excessive blocking, therefore an antiblock agent (5% w/w) was incorporated into the blend mixtures prior to extrusion.

Three layer coextruded films were also produced, with the mPE polyethylene as the middle layer. The ratio of mPE to LDPE in these films was the same as that in the blended films so all films were comparable.

The blended materials were also compounded using a four strand compounding die to produce well dispersed samples suitable for rheological analysis, described later.

TENSILE TESTING

All films were tested to BS 2782: Part 3 using an Instron 4411 universal tensile tester. Crosshead speed was set at 500 mm/min with a gauge length of 50 mm. For each film a minimum of ten samples were prepared and tested in both machine and transverse direction.

From the resultant data, values for break strength, % elongation and Young's modulus were obtained.

DSC ANALYSIS

Crystallinity of various films were determined using Differential Scanning Calorimetry. This was carried out using a Perkin Elmer DSC 6, over the temperature range 20 to 150°C with a scan rate of 10°C/min. Thermograms from first heating were used to determine crystallinity developed during processing. A value of 289.9 J/g was used to represent a 100% crystalline sample.[5]

RHEOLOGICAL ANALYSIS

Rheological data on the various polymer blends was obtained using a Rosand RH7 dual capillary rheometer. Each compounded blend was subjected to a range of shear rates ($10-10^3$ s^{-1}) at a constant temperature of 200°C.

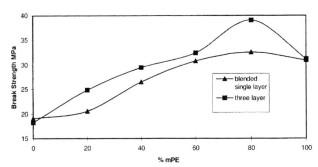

Figure 1. Comparison of break strength in MD for blended and coextruded films at haul of 3 m/min.

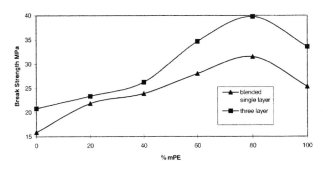

Figure 2. Comparison of break strength in TD for blended and coextruded films at haul off 3 m/min.

RESULTS AND DISCUSSION

Break strengths in the machine and transverse directions for both coextruded and blended films, manufactured at a haul off speed of 3 m/min, are shown in Figures 1 and 2. Generally, it is shown that as percentage mPE increases the break strength increases, up to a maximum of 80% mPE content in all cases. Note that the break strength of 100% mPE film was always less than that of films containing 80% mPE.

The results also show that coextruded films generally had considerably higher break strength than that of the blended films, for example at 20% mPE content the coextruded film is over 17% stronger than the monolayer blended film in the machine direction. In the transverse direction the coextruded films exhibit an even greater increase in break strength over the blended films as mPE content is increased, with the coextruded film being over 20% stronger than the blended film for 80% mPE. Martinez[11] suggested from optical microscopy studies, that row nucleation, induced through melt orientation, occurs at the interface between the layers in the coextruded

Table 1. Effect of haul off speed on the machine direction break strength for blended and coextruded films

mPE, %	Blended film break strength, MPa				Coextruded film break strength, MPa			
HO	3	6	9	12	3	6	9	12
0	19	24	24	32	18	23	28	34
20	20	22	25	25	25	24	20	24
40	26	26	24	25	29	22	22	18
60	31	25	27	24	32	27	20	20
80	33	22	27		39	28	24	18
100	31	21			34			

film. This enhanced nucleation could account for the greater break strengths observed in the coextruded films.

The effect of haul off speed on break strength in the machine direction for blended and coextruded films is shown in Table 1. For 100% LDPE it is evident that break strength increases with increasing haul off. As mPE is incorporated into the film, however, break strength decreases with progressively higher haul off rates. This becomes evident at 60% mPE for the blended film and 40% mPE for the coextruded film. This would suggest that molecular orientation increased up to these processing conditions, but further increase in haul off resulted in weakening the morphological structure of the film. This is confirmed by significant reduction in elongation for these films, for example for 60% mPE blended film elongation falls from 833% at 3 m/min to 254% at 12 m/min. Note that the results for 100% mPE blended film at 9 m/min and 12 m/min are not included in the data. This was due to severe bubble instability at these haul off rates therefore no samples were taken. Also at 80% mPE for the blended film no value of break strength for 12 m/min haul off is included. This is because that while it was possible to manufacture this film, excessive blocking of the lay flat film occurred (even with 5% antiblock agent added), making it impossible to separate. This again was the case 100% mPE coextruded film at haul off 12 m/min.

The effect of incorporating mPE on the elongation of the films is shown in Figures 3 and 4. The results show considerable increase in elongation with increasing mPE content, with the greatest change in elongation being recorded in the machine direction, around a 45% increase (approximately 500% to 900% for both the blended and coextruded films). Addition of 20% mPE resulted in a 16% increase in elongation for the blended films and over 36% increase in elongation for the coextruded films in the machine direction. Again, it is interesting to note that although the highest elongations were recorded for films containing 80% mPE (900%), a decrease in elongation is evident for 100% mPE films. In almost all cases the recorded elongation's for coextruded films were higher than that of the blended films.

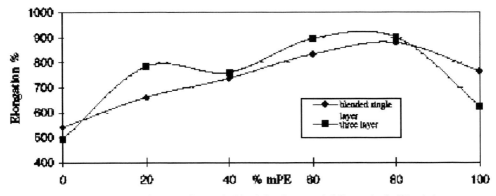

Figure 3. Comparison of percentage elongation in MD for blended and coextruded films at haul off 3 m/min.

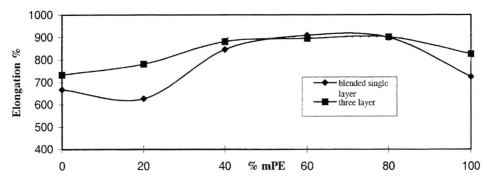

Figure 4. Comparison of percentage elongation in TD for blended and coextruded films at haul off 3 m/min.

The effect of increasing the mPE content on the Young's modulus of the various films was also studied. The results, in Table 1, show that this modulus generally decreases with increasing mPE content for all films, with a greater decrease being apparent for the blended films.

The increase in elongation and decrease in modulus with increasing mPE content would tend to suggest changes in the crystallinities of the films. DSC analysis on the blended films show this to be the case.

Table 2: Young's modulus for blended and coextruded films in MD and TD

mPE, %	Blended film, MPa		Coextruded film, MPa	
	MD	TD	MD	TD
0	176	162	137	143
20	134	136	129	130
40	118	109	105	119
60	89	101	95	103
80	74	79	68	80
100	55	56	74	80

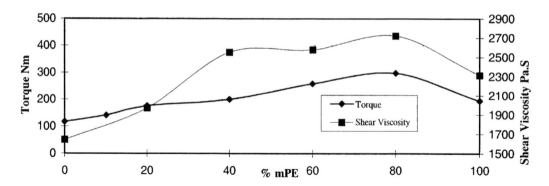

Figure 5. Comparison of torque and shear viscosity data for blended films at shear rate 10 s⁻¹.

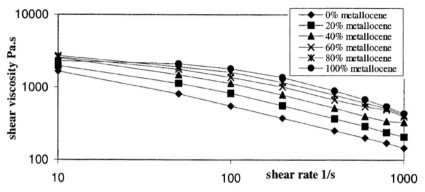

Figure 6. Shear viscosity vs. shear rate for mPE/LDPE blends.

Table 2 shows a significant difference in crystallinity was recorded for 100% LDPE (36%) and 100% mPE (24%). The results also show considerable decrease in crystallinity in films containing even small percentages of mPE, for example incorporation 20% mPE results in a drop of over 10% crystallinity. Recent studies[12] on α-olefin copolymers have shown that higher molecular weights will lead to a decrease in crystallinity. Since the molecular weight of mPE's are generally very high, this could account for the fall in recorded crystallinity . The thermograms for all blends showed only one peak at this cooling rate. The table also shows a decrease in melting point of the films as percentage mPE is increased.

During extrusion of the blended films torque measurements were recorded. Figure 5 shows an almost linear increase in torque with increasing mPE up to 80% mPE (290 Nm). However, extrusion of 100% mPE resulted in a smaller torque value. Following this observation, rheological studies on these blends were performed. The recorded shear viscosity data

Table 3. DSC results for blended films

mPE, %	Melting point, °C	ΔH, J/g	Crystallinity, %
0	110.7	105.04	36.23
20	110.1	72.68	25.07
40	108.3	67.25	23.32
60	106.3	64.67	22.31
80	101.2	73.62	25.54
100	97.8	71.23	24.57

over a range of shear rates for the blends is shown in Figure 6. Generally shear viscosity decreases with increasing shear rate. Note that at low shear rates higher shear viscosity's were recorded for 60% and 80% mPE than 100% mPE. As shear rate was increased however, 100% mPE exhibited the highest shear viscosity, and does not show as linear a decreases in shear viscosity with increasing shear rate as the other blends. This would indicate that mPE is less shear thinning than LDPE. Increasing the percentage of mPE will generally result in a higher shear viscosity.

Since the films for this study were manufactured at a relatively low shear rate, typically around 10 s^{-1} the rheological characteristics of the blends at this shear rate were of particular interest. This data is compared with the torque measurements in Figure 5. The rheological data follows the same trend as the torque data exhibiting a general increase in shear viscosity with increasing mPE content and confirms that the viscosity of 100% mPE is somewhat lower than an 80% blend as observed from the torque readings.

CONCLUSIONS

This present study would tend to indicate the following conclusions:
- Incorporation of mPE into an LDPE film will result in an increase in tensile properties, both for the monolayer blended film and the three layer coextruded film.
- Generally the coextruded films exhibit higher break strengths than blended films for the same percentage mPE
- Blending of even relatively small amounts of mPE into a LDPE film will lead to a significant decrease in crystallinity.
- Shear viscosity of the blends increases with increasing mPE content and 100% mPE was shown to be less shear thinning than LDPE.

ACKNOWLEDGMENTS

The authors would like to thank Jordan Plastics Ltd., Portadown, N. Ireland, for providing the materials for this study.

REFERENCES

1   http://www.modplas.com/month0597/film05.htm.

2   Kaminsky, *Macromol. Chem. Phys.*, **197**, 3907-3945 (1996).
3   UK Freezer Film Symposium, Birmingham, November 12, 1996.
4   Kim, Y. J., Chung, C., Lai, S.Y., Hyun, K. S., *Korean J. of Chem. Eng.*, **13**(3), 294-303 (1996).
5   Ramesh, N.S., Malwitz, N., Antec 1954-1961 (1997)
6   McNally, G.M., Bermingham, C., Murphy, W.R., *Trans IChem.E*, Vol **71**, Part A, 223-231 (1993).
7   Hill, M.J., Barham, P.J., Keller, A., *Polymer*, **33**(12), 2530-2541 (1992).
8   Iragorri, J.1.,Rego, J.M., Katime, I., *Polymer*, **33**(3), 461-467 (1992).
9   Norton, D.R., Keller, A., *J. of Mat. Sci.*, **19**, 447-456 (1984).
10   Datta, N.K., Birley, A.W., *Plastics and Rubber Processing Applications*, **3**, 237-242 (1983).
11   Martinez, F., Mazuera, G.E, *Plastic Film Technology*, **2**, 120-138 (1993).
12   Kennedy, M. A., Peacock, A. J., Failla, M. D., Lucas, J. C., Mandelkern, L., *Macromolecules*, **28**, 1407-1421 (1995).

Trade-offs in Blown Film Processing-Structure-Property Behavior of LLDPE Type Resins from Chromium, Metallocene and Ziegler-Natta Catalysts

Ashish M. Sukhadia
Phillips Petroleum Company, 94-G PRC, Bartlesville, OK 74004

INTRODUCTION

Polyethylene (PE) is most versatile polymer available today. It is inherently "simple" in it's chemical make-up, consisting essentially of carbon and hydrogen atoms bonded together co-valently. However, the wide variety of molecular chain architectures that have been made possible in the last 50 years or so have given rise to a tremendous number of PE resin-end user performance combinations.

Within the family of PE resins, linear low density PE (LLDPE) resins have found extensive applications in the areas of food packaging (e.g. bakery, candy, frozen foods, ice-bags, etc.), non-food packaging (e.g. garment bags, heavy duty sacks, stretch and shrink wrap, etc.) and non-packaging (e.g. industrial can liners, agricultural films, medical, trash bags etc.). These numerous applications demand a vast range of application performance benefits that can rarely be met by one resin alone. This has essentially resulted in the development of an array of catalysts, processes and end use resins over the last five decades. However, a comprehensive review of these variations is well beyond the scope of this paper.

Chromium catalyst PE resins are typified by a high molecular weight (M_w), broad molecular weight distribution (MWD) and often very low levels of long chain branching (LCB).[1,2] Ziegler-Natta (Z-N) resins are typically narrower in MWD than chromium resins and contain no LCB.[1,2] Metallocene resins are typically even narrower in MWD than Z-N resins and may[3] or may not[4,5] contain any LCB depending on the catalyst and process employed. Thus it is clear that these three catalysts viz. chromium, Z-N and metallocene can yield PE resins with widely different molecular and rheological properties. In this paper, we examine the blown film processing - structure - property behavior and trade-offs with three different LLDPE resins from each catalyst mentioned above.

EXPERIMENTAL
BLOWN FILM RESINS

Three LLDPE type resins made using chromium, metallocene and Z-N catalysts were used in this study. All three resins are hexene copolymers. For the sake of brevity, the chromium, metallocene and Z-N resins will be designated as Resins C, M and Z, respectively. The basic characterization data of these three resins is shown in Table 1. Note that resins M and Z are nominally 1 MI while resin C is 0.22 MI. This particular chromium catalyst resin C was chosen since it is a commercial resin and is therefore readily available. This difference in MI is not expected to affect the general interpretations and basic theme of this paper.

BLOWN FILM EVALUATIONS

The film blowing experiments were conducted on a laboratory scale blown film line. The line consists of a 38 mm (1.5 inch) diameter single screw Davis Standard extruder (L/D=24; 2.2:1 compression ratio) fitted with a barrier screw with a Maddock mixing section at the end. The die used was 100 min (4 inch) in diameter and was fitted with a Dual Lip Air Ring using ambient cooling air. The die gap was 1.52 min (0.060 inch). The film blowing was done using typical LLDPE conditions as follows: 2.5:1 blow up ratio (BUR), "in-pocket" bubble and 190°C extruder and die set temperatures. Films of three different gages viz. 25 micron (1 mil), 75 micron (3 mil) and 125 micron (5 mil) were made by varying the take-up speed while keeping the output rate constant for all resins and film gages. The evaluations were made at an output rate of 27.2 kg/hr (60 lb/hr) which was achieved by varying the extruder screw speed as needed. The film blowing conditions were chosen since the film properties so obtained scale directly with those from larger commercial scale film blowing operations.[6]

BLOWN FILM PROPERTIES

The film properties were measured as follows:
1   Dart impact strength - ASTM D- 1709 (method A)
2   Machine (MD) and transverse (TD) direction Elmendorf tear strengths
     - ASTM D-1922

RHEOLOGY

Melt index (MI) was measured as per ASTM D-1238, Condition F (190°C, 2.16 kg). High load melt index (HLMI) was measured as per ASTM D-1238, Condition E (190°C, 21.6 kg).

The rheological data was obtained on a Rheometrics Mechanical Spectrometer (model RMS-800) using the time-temperature superposition procedure outlined elsewhere.[5] The

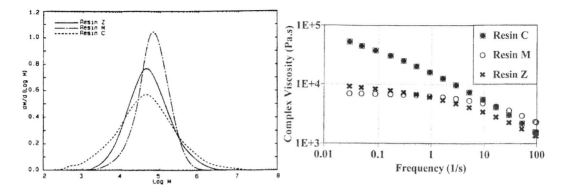

Figure 1. Gel permeation chromatrogram traces of LLDPE resin C, M, and Z.

Figure 2. Comparison of the complex viscosity of LLDPE resin C, M, and Z.

Table 1. Description of LLDPE resins C, M and Z

Property	Resin C	Resin M	Resin Z
MI, g/10 min	0.22	1.0	1.0
HLMI, g/10 min	18	17	37
HLMI/MI	82	17	37
Density, g/cc	0.922	0.918	0.921
Eta(0), Pa.s	2.87E5	7.18E3	1.15E4
Tau Relax, s	1.03	0.014	0.028
'A'	0.206	0.579	0.339
Ea, kJ/mol	33.9	23.3	23.9
Mw, kg/mol	219.22	101.08	122.04
Mn, kg/mol	10.22	39.58	25.55
MWD	21.5	2.6	4.8

master curve was fitted with a simplified three parameter Carreau-Yasuda (CY) empirical model to obtain the CY parameters viz. zero shear viscosity - Eta(0), characteristic maximum relaxation time - Tau Relax and 'A' parameter. Details of the significance and interpretation of these three parameters may be found elsewhere.[8,9]

Apparent extensional viscosity was experimentally obtained using the converging flow method proposed by Cogswell.[10] A two-part die consisting of a 2 mm diameter entry die with a 90° entry cone angle and a 2 mm die diameter land die (L/D=15) were fitted to the end of a 2.54 cm diameter extruder. The temperature profile used was 190°C flat across the extruder, adapter and dies.

RESULTS

We begin by examining the basic resin molecular and rheological characteristics for the three LLDPE type resins, C, M and Z. The gel permeation chromatograms (GPC) for the three res-

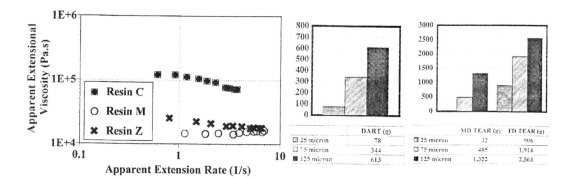

Figure 3. Apparent extensional viscosity vs. apparent extension rate for LLDPE resins C, M, and Z at 190°C via Cogswell's[10] converging flow method.

Figure 4. Effect of gage on the blown film properties of chrome catalyst LLDPE resin C.

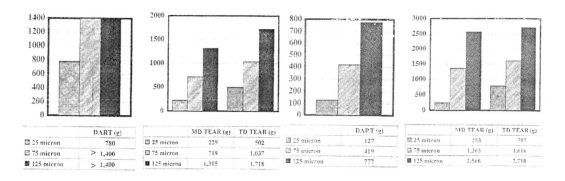

Figure 5. Effect of gage on the blown film properties of metallocene catalyst LLDPE resin M.

Figure 6. Effect of gage on the blown film properties of Ziegler-Natta catalyst LLDPE resin Z.

ins are shown in Figure 1 and the rheology (complex viscosity versus frequency) at 190°C in Figure 2. From Figure 1, it is clear that resin C is the broadest and resin M the narrowest in MWD, with resin Z in between. The molecular weight data is tabulated in Table 1. The rheology data in Figure 2 shows that the low-shear viscosity is highest for resin C and lowest for resin M. The high-shear viscosity, on the other hand, is lowest for resin Z and highest for resin M. The viscosity data is also summarized in Table 1 through the use of the Carreau-Yasuda model fit parameters as discussed in the experimental section.

The apparent extensional viscosity as a function of apparent extension rate, obtained using Cogswell's converging flow method,[10] is shown in Figure 3 for all three resins.

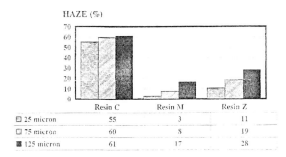

Figure 7. Effect of resin type and gage on the % haze of blown films.

The basic blown film properties viz. dart impact strength, MD and TD tear strengths for resins C, M and Z are shown in Figures 4, 5 and 6, respectively, and the % haze data for all resins and gages in Figure 7.

Lastly, the blown film processability data (with regard to extruder variables) is shown in Figures 8a-8c wherein the flow rate, pressure drop and motor load are shown as a function of screw RPM in Figures 8a, 8b and 8c, respectively.

ANALYSIS AND DISCUSSION

The detailed molecular and rheological data in Table 1 shows that the chromium catalyst resin C is the broadest in MWD, followed by the Z-N resin Z, followed by the metallocene catalyst resin M which is the narrowest in MWD. In addition, the weight average molecular weight, M_w, follows that same trend. The flow activation energy, E_a, further indicates that the chromium resin C has some (low) level of LCB whereas the Z-N resin Z and metallocene resin M do not.[5] The rheological CY parameters (see experimental section for details) show that resin C has the highest zero shear viscosity, Eta(0) followed by resin Z and then resin M. The characteristic melt relaxation time, Tau Relax, is the highest for resin C, followed by resin Z and then resin M. The CY parameter W, which is inversely related to the breadth of the relaxation time distribution shows, as expected, that resin C has the broadest distribution of relaxation times followed by resin Z and M.

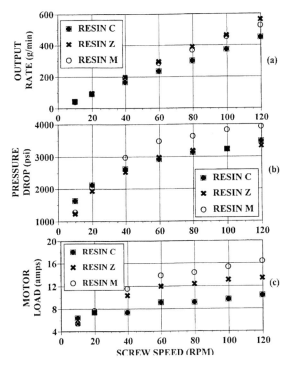

Figure 8. Extruder processability data of LLDPE resins C, M and Z. X axis is screw speed (RPM) in each case.

From Figure 4, the chromium resin C exhibits very low MD tear and dart impact strength at the lowest gage of 25 microns. Both these properties improve with increasing gage. However, it is remarkable that the increase in properties, particularly MD tear strength, is more than proportional to the change in gage. Going from 25 to 75 to 125 microns i.e., increasing film gage by a factor of 3 to a factor of 5, respectively, increases MD tear strength from 32 to 485 to 1,322 g which is an improvement of almost 16 times to 40 times, respectively, over the 25 micron film. The dart impact strength for resin C see similar, but much less dramatic, improvements with gage. The TD tear strength, on the other hand, increases less than proportionally with gage.

From Figure 5, the blown film data as a function of film gage for the metallocene resin M show very different trends than those observed above for resin C. First, the resin M exhibits very high impact strength even at 25 micron gage. At the 75 and 125 micron gages, the dart impact strength is in excess of 1,400 g which was the highest impact strength measurable with the current equipment. The MD tear strength at 25 micron is better for resin M than resin C while the TD tear strength is lower. At higher gages, resin M exhibits a change in MD tear that is roughly proportional to the change in gage while the TD tear strength increases less than proportionally with gage. It is interesting to note that at the 125 micron gage the MD tear strength of resin C and M are comparable while the TD tear strength of resin C is much higher than resin M.

From Figure 6, the blown film data as a function of film gage for the Z-N resin Z shows yet different behavior. The properties at 25 micron gage show the tear properties of resin Z to be equivalent to resin M and better than resin C. Dart impact strength of resin Z is higher than resin C but lower than resin M. At higher gages, the dart impact and MD tear strength increase more than proportionally while the TD tear strength increases less than proportionally with film gage. At higher gages, resin Z exhibits clearly superior MD tear strength compared to resins C and M.

At least some of these film properties and trends can be explained reasonably well on the basis of the molecular and rheological characteristics of these different polymers. The low MD tear strength of resin C, particularly at low gages, is attributed to the high degree of MD molecular orientation developed in the film during blowing. At low gages, i.e., high film line speeds, the process time (time for melt to go from die exit to freeze line height (FLH)) is short relative to the relaxation time of the polymer. Thus the polymer does not have a chance to relax much and a high degree of molecular orientation is "frozen-in". In addition, the high M_w and broad MWD of resin C, coupled with the fact that a small degree of LCB is also present in the resin, results in relatively high extensional viscosity for resin C compared to resins M and Z as shown in Figure 3. The higher extensional viscosity of resin C most likely results in a higher level of stress in the bubble prior to the FLH which further contributes to the generation of high MD orientation in the final film. In fact, wide angle x-ray scattering (WAXS) and infrared dichroism (IR) results (not presented here) confirm these postulations of very high molecular orientation along the MD in the 25 micron blown film from resin C. As the resin gage is increased, i.e., drawdown or line speed is decreased, there is a simultaneous increase in the process time (which allows for greater molecular relaxation) and a decrease in stress in the bubble due to lower drawdown stresses. These two effects combine to decrease the final orientation developed and retained in the film, thereby increasing the MD and TD tear properties of resin C.

It is of interest to note that the positive effect of thicker gage on tear strengths, particularly MD tear, is significantly greater for resin C compared to resins M or Z as noted earlier. This is because the rheological properties of resin C viz. zero shear viscosity, melt relaxation time and extensional viscosity are all considerably higher for resin C than either resin M or Z as discussed earlier. Thus any processing changes (such as lower line speeds) during film blowing that provide for added molecular relaxation to occur will affect the orientation state and solid state morphology of resin C to a much greater degree than resins M or Z as observed. As a final note on the tear strength properties, it should be added that the Z-N resin Z exhibits far superior MD and TD tear strengths and balance at all gages compared to resins C and M.

With regard to the impact strengths of these resins at the various gages, it is very clear that the metallocene resin M exhibits clearly superior toughness compared to resins C and Z. In fact, from Figures 3, 4 and 5, we observe that the dart impact strength of 25 micron resin M film is higher than even the 125 micron film properties of either resin C or Z. This is truly a remarkable difference in solid state properties. Metallocene resins are characterized by narrow MWD and homogenous short chain branching distribution (SCBD).[3-5] This homogenous SCBD is more efficient at forming a greater number of tie-molecules because of the fact that the evenly distributed comonomer (across the entire MWD[5]) disrupts the process of regular chain-folding into lamella more effectively. The benefits of increased tie-molecule concentration on the solid state properties[11-13] and toughness properties[14,15] of polyethylene have been discussed many times by other researchers.

In Figure 7, the % haze for all the blown films is shown. The metallocene resin M is far superior in clarity (as determined by low % haze) at all gages compared to either resin C or Z. The reasons for the superior clarity of this family of metallocene resins has been discussed elsewhere.[5] It is perhaps worthwhile to point out here that the high haze for resin C at all gages was determined (by applying a light coating of silicone oil of similar refractive index on both sides) to be largely due to fine-scale surface roughness (i.e., external haze not internal) . The surface roughness is postulated to be caused at the die exit due to the high polymer melt elasticity of this resin. On the other hand, the higher haze of resin Z compared to resin M is thought to be due to surface roughness caused by growth and aggregation of crystallites at or near the film surfaces. This in turn is postulated to occur in resin Z to a larger extent than resin M due to the more heterogeneous SCBD in the former.

Lastly, we examine and compare the processability behavior of these three resin types as shown in Figures 8a-8c. Figure 8a shows that the higher M_w, broader MWD resin C exhibits lower output for a given screw speed compared to resins Z and M. On the other hand, both the pressure drop and motor load for resin C are lower than resin M or Z. Resin M with a very narrow MWD and no LCB requires considerably higher pressure drops and motor loads, particularly at the higher screw speeds. These results are in good general agreement with those made by another group.[16]

For the sake of completeness, a few other processability observations are noted here. Resin C has much better melt strength compared to resin M and Z, as may be expected based on the extensional viscosity data in Figure 3, resulting in much better bubble stability in film blowing. In fact, resin C can be processed at both LLDPE ("in-pocket") as well as HDPE (high stalk) conditions at commercial film blowing rates. Resins M and Z simply do not possess the required melt strength for high stalk applications. Furthermore, because of the considerable differences in melt rheology and hence bubble stability among the three resins, the maximum output rates possible for resins Z and M are lower by about 15% and 30%, respectively, compared to resin C under the same film blowing conditions. In addition, resin C can also be processed at much higher rates without sharkskin melt fracture (SSMF). The onset of SSMF for resins Z and M occur at output rates lower by about 40% and 70%, respectively, compared to resin C. Resin C thus requires no process aid, resin Z may depending on the exact processing conditions and resin M does require processing aid in typical commercial film blowing situations.

CONCLUSIONS

The trade-offs in the blown film processing-structure-property behavior of three LLDPE type resins made using chromium, metallocene and Ziegler-Natta catalysts were examined in detail in this study. The LLDPE resin made from chromium catalyst had the highest M_w, MWD, zero shear viscosity, melt relaxation time and extensional viscosity. The LLDPE resin from Ziegler-Natta catalyst was intermediate while the LLDPE resin from the metallocene catalyst was the lowest in terms of these same attributes.

The chromium catalyst resin provided the best processability in film blowing exhibiting excellent bubble stability, high output rates, low extruder pressures and motor loads and high output rates for the onset of sharkskin melt fracture. In other words, the resin clearly offered the widest processing window. The blown film properties exhibited low dart impact and MD tear strengths, especially at low gages. However, these properties improved dramatically and more than proportionally with increasing film gage. The films exhibited poor optical properties at all gages as evidenced by high % haze. This chromium catalyst resin (and other resins similar in molecular and rheological characteristics) are thus best suited for non-clarity, high gage applications where high output rates and a broad processing window are desired.

The metallocene LLDPE resin had the narrowest processing window exhibiting the lowest bubble stability, highest extruder pressures and motor loads and lowest output rate for the onset of sharkskin melt fracture. The blown films from this resin, however, had the best optical properties as evidenced by very low % haze. The blown film properties increased roughly proportionally with film gage. The impact strengths of these films were excellent at all gages with the 25 micron metallocene LLDPE film notably exhibiting higher dart impact strength than 125 micron films from either the chromium or Ziegler-Natta resins. This metallocene resin (and other similar resins) are thus best suited in applications where high clarity and optical properties, excellent toughness and good tear strength are desired, at the possible expense of processability relative to other LLDPE resins.

The Ziegler-Natta resin had a processing window intermediate to the chromium and metallocene resins. The blown films exhibited reasonably good optical properties which were much better than the chromium resin but much worse than the metallocene resin. The blown film properties at all gages exhibited very good MD and TD tear strengths. Dart impact strength, however, was much poorer compared to the metallocene resin. This resin (and other similar resins) are thus best suited in applications where medium clarity or optical properties, medium impact or toughness, high MD and TD tear strengths and good processability are desired.

As a final note, it is important, worthwhile and necessary to clarify that many other blown film end use properties and attributes e.g. blocking, gas and water transmission rates, hexane extractables (FDA approval), sealing strengths and initiation temperatures, tensile properties, etc. were not included in this study, primarily due to space limitations. Thus the comparisons and tradeoffs discussed here, while still entirely valid and legitimate in their own regard, should not be considered as absolute but rather a starting point at differentiating these resin classes. The ultimate choice of a resin for a given application will depend on the complete balance and tradeoffs it offers in terms of both performance and price.

ACKNOWLEDGMENTS

The author would like to sincerely thank the following for their help: D. C. Rohlfing, T. W. Johnson, D. E. Higbee, J. D. Stark. Also, the support of Phillips Petroleum Company in publishing this work is gratefully acknowledged.

REFERENCES

1   Ser van der Ven, **Polypropylene and other Polyolefins: Polymerization and Characterization**, *Elsevier Science Publishers*, Amsterdam (1990).
2   M. P. McDaniel, *Adv. Catal.*, **33**, 47-98 (1985).
3   T. I. Butler, S. Y. Lai and R. Patel, *J. Plast. Film Sheeting*, **10**, 248 (1994).
4   D.J. Michiels, Worldwide Metallocene Conference, Metcon'94, May 25-27, Houston, TX, USA (1994).
5   A. M. Sukhadia, *J. Plast. Film Sheeting*, **14**(1), 54(1998).
6   A. M. Sukhadia, *J. Plast. Film Sheeting*, **10**(3), 213 (1994).
7   J. D. Ferry (ed.), **Viscoelastic Properties of Polymers**, 3rd edn., *John Wiley and Sons*, New York, 1980.
8   C. A. Hieber and H. H. Chiang, *Rheol. Acta*, **28**, 321 (1989).
9   C. A. Hieber and H. H. Chiang, *Polym. Eng. Sci.*, **32**, 931(1992).
10 F. N. Cogswell, *Trans. Soc. Rheol.*, **16**(3), 383 (1972).
11 A. Lustiger and N. Ishikawa, *J. Polym. Sci. Polym. Phys.*, **29**, 1047 (199 1).
12 A. Lustiger and R. L. Markham, *Polymer*, **24**, 1647 (1983).
13 N. Brown and I. M. Ward., *J. Mater. Sci.*, **18**, 1405 (1983).
14 T. M. Liu and W. E. Baker, *Polym. Eng. Sci.*, **32**(14), 944(1992)
15 K. C. Koch and B. V. Volkenburgh, SPE ANTEC Conf. Proc., 1879 (1996).
16 R. E. Christensen and C. Y. Cheng, *Plast. Eng.*, June, 31 (1991).

Transparent Metallocene-Polypropylene for Injection Molding

J. Rösch
BASF AG, Central Polymer Research Laboratory, Ludwigshafen, Germany
J. R. Grasmeder
TARGOR GmbH, Application Technology, Ludwigshafen, Germany

INTRODUCTION

PP is among the fastest growing plastics with annual growth rates of up to 6%. The versatility of polypropylene (PP) continues to expand as resin producers implement improvements in catalyst and process technologies and refine post reactor modification. Recently, several approaches towards the production of polypropylene with improved clarity have become available in the marketplace. Transparent polypropylenes are gaining an ever widening acceptance for packaging applications. State of the art transparent polypropylenes challenge polystyrenes in many applications and offer the additional benefit of improved toughness.

BACKGROUND

The intrinsic reason for the turbidity of conventional polypropylene is its semicrystalline state, unlike long-established transparent and amorphous polymers like polystyrene. Light scattering, occurring at spherulite boundaries, gives rise to the opaque appearance of most semicrystalline polymers. The key to improved clarity is the control of the number and size of the spherulites. This can be done with different approaches described in the following.

One of the earliest attempts to improve the clarity of polypropylene was to use low levels of external stereomodifiers during the catalytic polymerization of propylene. This yielded products with decreased stereoregularity and therefore decreased ability to crystallize. These polypropylenes, rich in chain defects, reduced the overall crystallinity and also produced small spherulites with low perfection. Both effects improved clarity. However, these products did not succeed in the market place, because the reduced stereoregularity was inevitably accompanied by high levels of xylene solubles. These high levels of extractables rendered the products sticky and were unacceptable in most applications.

One of the most popular approaches towards transparent polypropylene is the post-reactor modification with selected nucleating agents. First generation nucleating agents like sodium benzoate produce a spherulitic morphology with spherulite diameters around 1 μm. Unfortunately this size is comparable to the wavelength of visible light and therefore produces a maximum scattering intensity according to the Mie-theory. For even smaller spherulites, a higher number of nuclei would be required. In sodium benzoate and other salt-type nucleating agents, minimum particle size and dispersion are limited by the ability to grind and disperse such non-fusable salts.

Additionally a small number of highly birefringent β-spherulites can often be found in samples of PP, especially below the surface layers. These highly birefringent inclusions also contribute to increased turbidity.

EXPERIMENTAL

Ziegler-Natta PP's were standard grades from BASF Novolen grade range. Metallocene PP's were TARGOR's Metocene development grades.

Nucleating agents were mixed with polypropylene powders by standard compounding equipment (Werner &Pfleiderer ZSK 30).

The sol-gel transition temperatures of PP-PDTS (PDTS=para-ditoluylidene sorbitol) mixtures at different PDTS levels were measured by oscillatory shear rheometry (Rheometrics RMS 800, temperature sweeps).

Transmission electron microscopy was performed on a Zeiss CEM 902 using ruthenium tetroxide stained and microtomed thin sections.

Haze values were determined according to ASTM D 1003 on 1 mm thick injection-molded samples. Transparency is represented as "100 - Haze".

RESULTS AND DISCUSSION

To increase clarity, an effective nucleating agent has to prevent the formation of highly birefringent β-modification and most important to reduce spherulite size to a minimum. To do so, the number of nuclei has to be increased drastically. Therefore dispersion of nucleating agents on a nanometer scale has to be accomplished. However, on a nanometer scale, dispersion cannot be achieved any longer by mixing operations. Thus "nanodispersion" nucleating agents require compounds to be soluble in a PP matrix. To be effective as a nucleating agent, however, there has to be a liquid(PP)/solid(nucleating agent)-interface. To create this interface, "nanodispersion" nucleating agents have to phase-separate prior to crystallization of the PP matrix. How this phase separation together with "nanodispersion" can be accomplished by physical gelation has been investigated by TEM, WAXD and rheometry. The structural causes for this type of physical gelation mechanism have been investigated in particular for para-ditoluylidene sorbitol (PDTS) as nucleating agent in metallocene-PP.

When cooled from homogeneous solution to the sol-gel transition temperature, spontaneous gelation of PDTS occurs. From colloid science it is a well-known fact that certain low molecular weight compounds are able to form macroscopic gels via a self-assembly process. In the case of PDTS, molecular modelling indicates that this self assembly proceeds by hydrogen bonding and stacking of phenyl rings (van der Waals attraction) and creates intermeshing fibrils of solid PDTS. This sol gel transition is thermodynamically favorable yielding 26 kcal/mol. This filament network can be pictured by transmission electron microscopy and shows PDTS fibrils of approximately 10 nm diameter. The calculated surface area of 350 m^2/ml for such a filament network proves the excellent state of dispersion attainable.

WAXD of the crystalline starting material and and freeze-dried PDTS-gels prepared from benzene-solution show the same crystal structure. However, size and perfection of the gel-crystals are much lower. Analysis of line width yields a crystal size of 35 nm in the crystalline starting material and a crystal size of 9 nm in the gel. This reflects well the nanofibrillar appearance seen in TEM. Using VR-INDEX software to resolve the crystal structure yields an orthorhombic unit cell with a = 20,21 Å b = 5,53 Å and c = 17,51 Å containing 4 PDTS molecules. The short b-axis indicates primitive stacking of the PDTS-molecules as proposed by molecular modelling.

These PDTS-fibrils now provide strings of nucleation sites leading to drastically reduced spherulite sizes and improved transparency.

The onset of gelation of PDTS in PP can be monitored by rheology. In cooling experiments storage modulus G, jumps up by more than a decade when the gelation of PDTS occurs. Of course the sol gel transition temperature is concentration dependent (Figure 1). In the region of commercial interest, lying around 0,1 and 0,3% PDTS the sol gel transition temperature lies between 160 and 180°C. This transition temperature is well above the onset of crystallization at 135-140°C. The mode of action of the benzylidene sorbitol type nucleating agents can therefore be summarized as follows: At typical processing temperatures, PDTS dissolves homogeneously in the PP melt. On cooling at a temperature around 180°C, self-assembly of PDTS molecules creates a network of intermeshing nanofibrils. These fibrils serve as highly dispersed nucleation sites for PP crystallization. The multitude of nucleation sites effectively limits spherulite growth to dimensions well beyond 1 μm and yields greatly improved clarity.

Furthermore the increased crystallization temperature favors the development of the thermodynamically more stable and more transparent γ-modification (Figure 2). Growth of β-spherulites is therefore effectively suppressed.

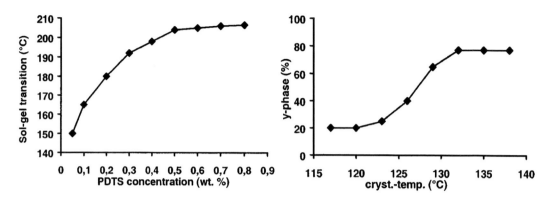

Figure 1. Concentration dependence of PDTS gelation temperature in PP

Figure 2. Dependence of γ-phase content on crystallization temperature.

Table 1. Characteristics of metallocene PP compared with Ziegler-Natta PP

Property	Unit	Metallocene homopolymer	Ziegler-Natta homopolymer
DSC melting point	^{o}C	148	162
Melting enthalpy	J/g	97	105
M_w/M_n	-	1.9	3.3-5
Xylene solubles	wt%	0.4	1.5-5
Chlorine content	ppm	trace	20

It is only recently with the advent of metallocene catalysts that PP homo- and copoly-mers with uniform molecular and compositional distribution have become available.[3] Selected properties of Z-N PP and metallocene PP are compared in Table 1. The homoge-neous composition of metallocene PP homopolymers synergistically works together with third generation nucleating agents to achieve a level of clarity that was previously only thought possible in random copolymers, while retaining the high stiffness of PP homopolymers. Figure 3 shows the different correlation of transparency and stiffness for Z-N PP and metallocene PP homo- and random-copolymers containing equivalent amounts of clarifier.

One of the reasons for the improved stiffness at constant transparency of metallocene PP over conventional Z-N PP is the compositional homogeneity of metallocene PP. High levels of xylene solubles occurring in conventional Z-N-random-PP act as undesirable plasticizers and decrease stiffness. Figure 4 shows this increase in xylene solubles for Ziegler-Natta random-PP with decreasing melting point (i.e., higher comonomer content). In contrast

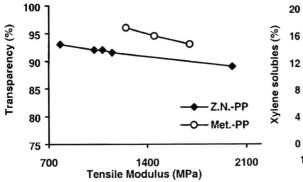
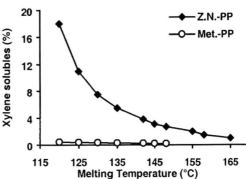

Figure 3. Correlation of stiffness and transparency for Z-N and metallocene PP.

Figure 4. Xylene solubles of Z-N and metallocene PP with respect to their melting point.

Table 2. Comparison of transparent PP with some other transparent polymers

Property	Unit	Test	PP	PS	SBS	PET
Density	g/cm^3	ISO 1183	0.9	1.05	1.0-1.2	1.2-1.3
Tensile modulus	MPa	ISO 527	800-2000	3100-3300	1100-1900	1900-2200
Charpy impact @ 23°C	kJ/m^2	ISO 179/1eU	70-NB	10-28	25-NB	80-NB
Notch Charpy impact @ 23°C	kJ/m^2	ISO 179/1eA	2-7	2-4	2-5	3-6
HDT (B)	°C	ISO 75	60-100	73-98	70-80	65-75

metallocene catalysts incorporate comonomers homogeneously for all chain lengths without concomitant formation of plasticizing xylene solubles. Metallocene PP of such homogeneous composition now forms very small, equally sized PP-crystallites with a high content of γ-modification. As a result, transparency/stiffness balance of metallocene homo- and copolymers is unsurpassed.

As an additional benefit of low xylene solubles organoleptic properties of metallocene-PP are superior to conventional random-PP grades. Compared to PS (Table 2) transparent metallocene-PP grades also offer improved impact strength. Thus the combination of high stiffness and transparency, good toughness and organoleptic properties and PP's ability to be hot filled make metallocene-PP the material of choice for transparent packaging applications.

CONCLUSIONS

The reasons for the unsurpassed transparency of metallocene-PP in injection molding applications has been investigated. The excellent transparency of metallocene-PP can be traced back to two causes. PDTS based nucleating agents achieve a "nanodispersion" of nucleation sites by a physical gelation mechanism that is related to the peculiar structure of the benzylidene sorbitols. Such highly dispersed nucleation sites drastically reduce spherulite size and increase the content of γ-modification both resulting in dramatically improved transparency. This nucleation mechanism synergistically works together with the excellent compositional homogeneity of metallocene homo- and random copolymers of propylene. This compositional homogeneity further improves transparency and stiffness. Transparent metallocene PP will therefore continue to challenge PS in transparent packaging applications especially when higher toughness is required.

ACKNOWLEDGMENT

The work and the results presented in this paper have been funded and commissioned by TARGOR GmbH.

REFERENCES

1   R. D. Leaversuch, *Modern Plastics Intl.*,Vol. **26**, No. 12, p. 54 - 57 (1996).
2   J. R. Grasmeder, Proceedings of Metallocenes '97 Europe, Düsseldorf, April 8 - 9, 1997.
3   W. Bidell et al., Proceedings of MetCon '96, June 13, 1996.
4   Data from own measurement and G. Lindenschmidt & N. Niessner, Kunststoffe, 85, 1066 - 1074 (1995).

Equipment Design Considerations for Processing with Metallocene Resins

Andrew W. Christie
The Black Clawson Co., 46 North First Street Fulton, NY 13069

INTRODUCTION

Material suppliers are continuously improving and upgrading their offerings to enhance the performance of the finished product. Periodically dramatic changes occur which significantly alter the material performance and open new product opportunities. The metallocene catalyzed polyethylenes in the 1970's, adoption of the new materials into new and existing applications is initially slow. There are several reasons for this delay in acceptance including:

1. Misunderstandings of the new material capabilities
2. Risk associated with customer acceptance and balance of properties
3. Costs which are initially high
4. Processability on existing equipment

This paper will discuss the factors affecting processability of these new resins and the impacts on equipment design.

FACTORS AFFECTING PROCESSABILITY

In any manufacturing process there are three key ingredients that go into making a product. These are the raw materials, the tools or equipment required to manufacture, and the human interaction to control the tools and convert the raw material into the finished product. The finished product has some target specification in physical dimensions and performance characteristics with an associated tolerance on what deviation from the target is acceptable. The variability in each of the three key ingredients influences the ability to reach the target. The process of making the final product is the synergistic interaction between the three ingredients. To understand the role of the equipment design in processing metallocene resins we

Table 1. Polymer properties

Feed	Size & shape
	Uniformity
	C.O.F. external
	C.O.F. internal
	Yield strength
Polymer	Melting point
	Heat capacity
	Thermal conductivity
	Viscosity
	Elasticity
	Shear sensitivity
	Temperature sensitivity
	Thermo-mechanical stability
	Melt density
	Additives

Table 2. Operation variables

Feed regulation
Screw speed
Temperature set points
Screen pack selection
Back pressure level

Table 3. System properties

Power/torque
Length/diameter
Temperature control
Screw design
Mixer design
Screen changer design
Back pressure valve design*
Melt pump design*
Pipes - length/diameter
Stationary mixer design
Coextrusion adapter design
Die manifold design
Die shape adjustment
* where applicable

must first look at the interactions between all three components of the process. The following three tables list selected variables that have a significant impact on plastics processing.

The structure of the polymer will dictate the polymer properties. The metallocene catalyst technology allows tailoring of the polymer structure to obtain specific finished product properties. The primary structural elements considered are:

1 molecular weight (MW)
2 molecular weight distribution (MWD)
3 long chain branching (LCB)
4 short chain branching (SCB)
5 comonomer type and amount.

Adjusting these structural elements of the polymer also affects processing. Increasing the MW while improving toughness causes the polymer to be more viscous during processing yielding higher processing temperatures. A narrow MWD improves strength properties but again reduces shear thinning yielding higher processing temperatures. A broad MWD will provide higher melt strength and improved optical properties. A high level of SCB will reduce crystallinity and stiffness providing toughness, good optics, and increased permeability. LCB will increase melt elasticity and surface haze and reduce gloss. The power of the metallocene technology is it allows one to build on this combination of properties as required. The next step is to review how these polymer property characteristics interact with the equipment design.

EQUIPMENT DESIGN CONSIDERATIONS

We can separate the equipment into three basic areas for the purpose of this design discussion, the extruder, the transport system, and the forming system. The key elements in the extruder being the drive system, the feed screw with mixer, and the temperature control system.

The transport system would include all elements in the flow path between the extruder and the forming system. This could include screen changer, back pressure valve, transfer pipes, stationary mixer, and metering pump. The forming system may include the coextrusion combining adapter and the forming die. In most instances the new polymers will be introduced to enhance the performance of an existing product and will be processed on existing equipment.

It is recommended by the commercial suppliers of metallocene resins to process on equipment intended for processing linear low density resins. From the previous discussion on material properties it is apparent as you target higher MW and narrow MWD and reduced SCS the material will become more difficult to process. Due to the higher crystallinity, higher viscosity, and reduced shear thinning it will require more energy. On existing equipment this translates into more horsepower or higher torque with higher processing temperatures and operating pressures. It is not unusual to see an increased power requirement as high as 25%. If this drive power is not available the options are to increase the drive motor horsepower or re-gear to a lower screw speed. Care should be taken to maintain an adequate service factor on both the gear box and the screw root.

The extrusion feed screw and mixer design will control the melt quality during processing. It is impossible to have a single feed screw design that is optimized for all the resins it may potentially run. For this reason it is most important to consider all current and potential materials that will be run on the screw and the amount of time they will be run. If the new resins are only a small part of the product mix to be run on a particular machine more compromise on the performance on that resin can be expected. Some of the short comings of a particular screw design may be offset by the operating conditions such as an inverted temperature profile to reduce drive amps and keep the temperature low. Again the commercial suppliers of metallocene polyethylene recommend the use of a barrier screw design. Typically barrier screw designs achieve higher melting efficiencies with lower shear energy input. As the energy input to the melt is a linear function of the materials viscosity a material that exhibits reduced shear thinning will gain more energy during processing. This material will also develop more pressure in the screw, transport system, and die. The higher system pressure means increased residence time in the screw contributing to yet higher temperatures. When a variety of materials are run a back pressure valve may be used to improve the performance of a deep screw on less viscous materials by increasing the effective shear rates and residence times. If a particular machine will be dedicated to running one material an optimized screw design can be developed.

It is important to discuss mixing in regards to screw design. Current state of the art of screw designs incorporates a mixing element. Typically this will allow increased outputs with improved quality. These mixing devices will breakdown through shear input (disperse) additives or unmelt or systematically rearrange (distribute) the flow stream to improve melt homogeneity and temperature uniformity. Some mixing devices serve both functions. A long barrier in a screw design is a form of dispersive mixer. When processing at its design point no additional dispersive mixing should be required. This is significant as inclusion of a second high shear mixer will consume more energy and may tend to overheat the polymer. In addition, many high shear mixers create a considerable pressure drop. This can create a potentially dangerous situation as most production extruders only have pressure indication downstream of the mixer. Actual operating pressures in the extruder may be 1-2 thousand psi higher than indicated by the pressure transducer with a poor pumping high shear mixer. An optimized mixer design then for metallocene resins should have low to moderate shear, provide forward conveying to reduce pressure losses, and create multiple flow re arrangements.

The transfer system and the forming system each provide resistance to the polymer flow. The extruder and feed screw must generate the pressure to overcome this resistance at the flow rate desired. Changing to a metallocene resin will generally increase the operating pressure of the system. Typically 1/3 or more of the system pressure drop occurs in the final forming lands of the die. High resistance also occurs in the screen changer and back pressure valve. Changes through the transfer system that will improve the pressure performance include increase the screening area, increase pipe diameters, increase valve clearances, and reduce transport pipe lengths and turns. Note that these first three actions will also each increase the average residence time in the system.

The forming system offers the greatest opportunity to reduce the overall system pressure. Increasing the final die opening is frequently possible by a simple die adjustment. In other cases it maybe accomplished through minor modifications to the die. A die replacement is a simple if somewhat costly procedure.

SUMMARY

A new family of metallocene catalyzed polymers has been introduced offering improved performance properties. When targeting increased MW, narrow MWD, and reduced SCB these polymers will be more difficult to process. Increased extruder torque, low shear screw designs to keep melt temperatures down, as well as low pressure transport and forming systems will be required for optimum processing. Many existing systems will be adequate for limited processing at reduced outputs and using operation special operating techniques to keep temperatures and pressures down. Particular attention should be given to mixing devices to enhance performance and protect from excessive pressures.

ACKNOWLEDGMENTS

I would like to thank Dr. Harry Mavridas at Black Clawson-SANO for his assistance and background information. Also Dow and Exxon for their assistance with technical literature and materials for laboratory trials.

Product and Process Developments in the Nitrogen Autoclave Process for Polyolefin Foam Manufacture

D. E. Eaves and N. Witten
Zotefoams plc, Croydon, UK

INTRODUCTION

There are several processes used for the production of cross-linked polyolefin foams[1-3] of which the two best known are:

THE CONTINUOUS ROLL PROCESS

In which a polymer sheet containing blowing agent and crosslinking additives is extruded and subsequently passed through a series of ovens which effect both the crosslinking and the decomposition of the blowing agent to initiate the foaming process. Alternatively, the crosslinking agent can be omitted and crosslinking is then achieved using on-line electron beam irradiation, although the absorption characteristics of the irradiation limits sheet thickness and process speed. The blowing agent is normally azodicarbonamide (ADC), and the crosslinking agent dicumyl peroxide.

THE BLOCK FOAM PROCESS

This involves pre-compounding of polymer, blowing agent, and crosslinking additive before transferring to a heated, heavy duty press where, at a higher temperature, both the cross-linking reaction and blowing agent decomposition are initiated. Foams of density down to around 70 kg/m^3 may be produced by this means. A process variant involves a second stage expansion in a separate mold which can yield foams of density down to 25 kg/m^3.

Both processes are widely used and account for the majority of crosslinked polyolefin foam production. However these processes have some disadvantages, namely:

- The foams can carry a characteristic, often unpleasant, odor resulting from the peroxide, together with ADC decomposition products (which include ammonia).

- The foams, particularly block process materials, can suffer from density variations (and hence property variations) through the thickness.
- Residual solids from the blowing agent are left in the foams.
- The foam cell structure is non-uniform.
- Dimensional changes can occur post manufacture through diffusional exchange of blowing agent gases with air.
- Property anisotropy can result from non-uniform expansion during processing.

THE NITROGEN AUTOCLAVE PROCESS

This process, although not widely used, has been operated commercially for over 60 years (by Zotefoams plc, Croydon, UK). There are three main stages i.e.:

STAGE I - EXTRUSION

A solid sheet of polymer is extruded and crosslinked using peroxides or irradiation. The solid extrudate is cut to size with profile, surface quality, dimensions, and crosslink level all being regularly assessed. There is no blowing agent addition at this stage.

STAGE 2 - GASSING & NUCLEATION

The cut slabs are loaded into autoclaves typically rated up to 69 MPa (10000 psi). The temperature is raised above the softening point of the polymer and pure nitrogen (the blowing agent) is introduced at high pressure. The solubility of nitrogen in polyethylene at a given temperature is linearly related to gas pressure by Henry's Law and this relationship controls the amount of nitrogen dissolved in the polymer and hence the final foam density. A further factor is crosslinking, which not only improves the physical strength of the finished foam, but stabilizes the sheet at the gas uptake stage, preventing polymer flow. When complete saturation with nitrogen has been achieved, a rapid reduction in the pressure effects nucleation as the gas comes out of solution. This nucleation step is critical and is accurately controlled since variation of the rate of gas release affects the cell size of the finished foam. The autoclave is then cooled and slabs removed for transfer to stage 3. The product at this stage is some 30% greater in volume than the original

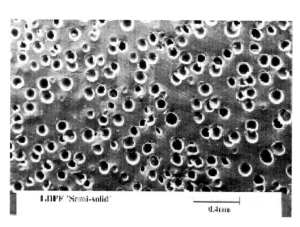

Figure 1. SEM micrograph showing the nucleated celles in a semi-solid (prior to expansion) sheet.

 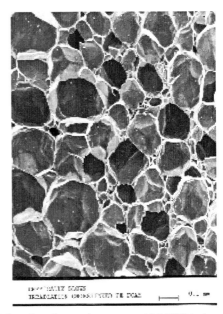

Figure 2. (left) SEM micrograph showing the cell structure of a LDPE foam from the autoclave process. (right) SEM micrograph showing the cell structure of a typical chemically blown, irradiation crosslinked PE foam.

and is essentially a foam of density ~ 720 kg/m^3. The size of the nucleated cells, which contain trapped nitrogen at the pressure of the autoclave, is of the order of 80 to 100 microns (see Figure 1).

STAGE 3 - FINAL FOAM EXPANSION

This final stage involves reheating the slabs under low pressure in a second autoclave to above the softening point of the polymer. At this point pressure is reduced and the slabs expand fully to a size dependent on the gas pressure in the cells and the original extruded solid size. Although the foam at this stage contains only nitrogen, eventually diffusion occurs exchanging the internal gas with air. However, since air is mostly nitrogen, there are no significant dimensional changes and no maturing is needed.

PROCESS & PRODUCT ADVANTAGES/DISADVANTAGES

Advantages are:
- Ability to control density and cell size by process, not formulation, variations.
- Environmental acceptability - the expansion agent is "borrowed" from the atmosphere, used to expand the foam, and then returned.
- Purity of product - there are no solid residues from blowing agent.

Table 1. Representative properties of autoclave processed vs. chemically blown foams (ASTM test methods)

Property	Units	Autoclave process	Chemical blow process	
			Block	Continuous roll
Polymer raw material	-	LDPE	LDPE	LDPE
Foam density	kg/m^3	42	48	41
Tensile strength	kPa	585	320	330
Elonagation @ break	%	146	120	80

Compression set (50%/22 h)
    2 h recovery
    24 h recovery
	%	
		8
		7
			
			27
			15
				
				17
				14

- Uniform cell structure, close to a pentagonal dodecahedron. Comparison with a chemically blown foam is shown in Figures 2(left) and 2(right). The structural differences are manifested in such foam properties as tensile strength and compression set, as shown in Table 1.

Disadvantages are:

- The restricted product form, i.e., thin block, typically 1 m x 2 m x 30 mm (although long rolls and thick blocks can be produced post manufacture by heat welding).
- The high capital cost of the process equipment. (Running costs however, tend to be lower than chemical blow processes, particularly at densities below 60 kg/m).

PRODUCT AND PROCESS DEVELOPMENTS FROM THE AUTOCLAVE PROCESS

HIGH DENSITY POLYETHYLENE (HDPE) FOAM

The process can provide a clear separation between extrusion to produce polymer slabs, subsequent crosslinking by off-line irradiation, and introduction of blowing agent (N_2) in the high pressure autoclave. This allows high temperatures to be used at extrusion (e.g. > 200°C) which in other processes would result in premature decomposition and reaction of crosslinking agent and/or blowing agent. Thus, foams based on HDPE (polymer density 0.96 g/cm^3, extrusion temperature around 200°C) can be produced using the autoclave process and are commercially available. Properties of these foams (designated HD) are shown in Table 2. Applications are mainly in areas requiring either cushioning performance at higher impact levels (e.g., lining for trauma protection in military helmets), or higher operating temperatures.

Table 2. Summary of the properties of HDPE and HD/LD blend foams produced using the nitrogen autoclave process

	Density	Compressive strength	Tensile strength	Elongation at break	Flexural modulus	Tear strength	Compression set
Test method	ISO 845: 1988	BS 4370: Pt 1: 3a:1988	ISO 1798: 1983	ISO 1798:1983	BS 4370: Pt 4:14: 1988	ISO 8067: 1991	ISO 1856:C:1980
Units	kg/m³	kPa	kPa	%	MPa	N/m	%
HD 30	30	115	825	55	3.6	1320	7
HD 60	60	300	2230	110	11.0	2725	11
HD 115	115	710	2860	110	23.0	6000	11
HL 34	34	90	700	50	2.2	1200	9
HL 47	47	130	800	50	5.2	2000	9
HL 79	79	275	1350	100	11.9	4500	9

MEDIUM DENSITY POLYETHYLENE (MDPE) FOAM

Low density polyethylene (LDPE) has a density of about 0.92 g/cm³. A range of polyethylene variants is available in the density range between LDPE and HDPE, and the properties (e.g. stiffness) of such materials can also be approached by blending polyethylenes of appropriate densities. Such materials have higher processing temperatures than LDPE, but may be foamed by the autoclave process using the irradiation crosslink route, as for HDPE above.

Such foams (designated HL) are also in commercial production, and properties are again shown in Table 2. The higher stiffness, compression strength, and creep resistance of these foams compared with LDPE foams, together with retention of the ability to recover from high compression strain (due to the uniform cell structure), gives these foams exceptional performance in cushion packaging applications.

Examples of cushion curves comparing HL foams with non-crosslinked (non-YL) packaging foams based on LDPE are shown in Figure 3. Performance of the non-XL LDPE foam is matched by an HL foam at half the non-XL foam density. A comparison of cushioning efficiency of different foam types can be made by calculating the weight efficiency factor:[4]

$$W_R = \frac{Foam\ density}{Energy\ absorbed\ per\ unit\ volume\ at\ C_{min}}$$

Table 3 shows such a comparison, again illustrating the effectiveness of HL foams in cushioning applications.

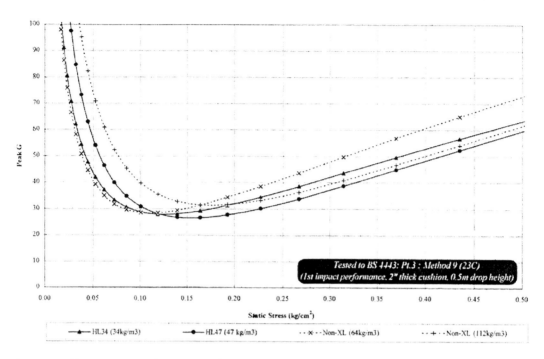

Figure 3. Cushion curves comparing HL grade foams from autoclave process with typical non-crosslinked polyethylene packaging grade foams. Note the density of the respective grades.

Table 3. Cushioning-weight efficiency factors (W_R) for HL, LD and non-cross-linked packaging foam grades

Foam grade	Density, kg/m³	1st drop W_R, 10^{-4}m^{-1}	1st drop mean	3rd drop W_R, 10^{-4}m^{-1}	3rd drop mean
LD	21	44	48	48	60
	29	48		63	
	44	54		66	
	69	46		62	
HL	33	26	26	44	41
	47	30		42	
	56	22		35	
Non-crosslinked	35	57	57	74	78
	54	57		76	
	100	57		84	

Table 4. Property comparison of nitrogen autoclave processed foams based on a metallocene-catalyzed polyethylene (MP) and standard density polyethylene (LD)

Property	Test method	Unit	Density=24 kg/m³		Density=33 kg/m³		Density=45 kg/m³	
			MP	LD	MP	LD	MP	LD
Comp. stress @ 25% strain	ISO 3386/1	kPa	42	35	50	40	61	50
Comp. stress @ 50% strain	ISO 3386/1	kPa	117	100	134	115	155	135
Tensile strength	ISO 1798	kPa	585	340	707	455	931	600
Elongation at break	ISO 1798	%	170	105	205	135	224	150
Tear strength	ISO 8067	N/m	896	410	1433	690	2335	1030
Comp. set	ISO 1856/C	%	8	8	5	7	4	7

METALLOCENE BASED FOAMS

Polyethylenes and other polyolefins produced using metallocene catalysts are now commercially available, and are claimed to have advantageous properties (e.g. film strength) compared with other polymers of comparable density and MFI although some disadvantages in processing have been reported.[5]

A range of metallocene catalyzed polyethylenes have been foamed using the autoclave process, and foam properties evaluated. It has been found that not all polymers are readily expanded by the process, nor are clear property advantages always demonstrated. However, selected grades have been found to foam without significant process difficulties and exceptional properties (tear strength, tensile strength, elongation and thermoformability) in comparison with LDPE based foams have been obtained. Properties of autoclave foams based on metallocene catalyzed polyethylene (polymer density 0.91-0.92 g/cm³, designated W) are shown in Table 4 (note the high tear and tensile strengths).

The versatility of metallocene catalysts also allows production of polyethylenes of lower polymer density and with a flexibility similar to EVA copolymers. These grades have also been evaluated using the autoclave process. Again, not all grades are readily "foamable" or give advantageous foam properties, but some grades can be foamed and show improved tensile and tear strength compared with EVA foams. A comparison of a flexible metallocene foam (using polymer density 0.87-0.89 g/cm³, designated MV) with EVA foam, both from the autoclave process, is shown in Table 5.

Table 5. Comparison of properties of foams produced by the nitrogen auto-clave process based on EVA copolymers with differing vinyl acetate contents and a low density (0-67-89 g/cm³) metallocene-catalyzed polyethylene elasto-mer (MV)

Property	Method	Unit	EVA (9% VA)	EVA (18% VA)	EVA (18% VA)	MV
Density	ISO 845	kg/m³	35	30	50	35
Comp. stress @ 25% strain	ISO 3386/1	kPa	35	35	40	37
Comp. stress @ 50% strain	ISO 3386/1	kPa	105	100	110	103
Tensile strength	ISO 1798	kPa	620	740	930	1050
Elongation at break	ISO 1798	%	200	200	220	246
Tear strength	ISO 8067	N/m	730	800	1300	1384
Compression set	ISO 1856	%	7	9	6	8

POLYPROPYLENE FOAM

In comparison with polyethylene, polypropylene has a temperature performance some 30°C higher owing to its higher softening point. The main problem in production of a crosslinked polypropylene foam, where superior performance at elevated temperatures can be antici-pated, has always been the different reactivity to peroxide or irradiation, which will crosslink polyethylene but will cause chain scission of polypropylene.

An approach to overcoming this problem is to use a blend of polypropylene copolymer and polyethylene (which also reduces processing temperature) together with a crosslink pro-moter such as tri-allyl cyanurate (TAC). However, the presence of polyethylene in the blend reduces temperature performance, and many crosslink promoters present health risks if used in open processes.

It has been found possible by using selected polypropylene copolymers to achieve a pro-cessing window within which the polymer slabs have sufficient stability to retain their shape during autoclave gassing, and expand subsequently without collapse. This process window is retained by the foam and allows the foam to be thermoformed similarly to a crosslinked foam. Since no polyethylene is incorporated, good performance at elevated temperature is assured, and the absence of covalent crosslinks together with a 100% polypropylene system, result in total recyclability if temperatures are raised above the process window.

This foam, (designated PA) has been developed over the last few years and is now com-mercially available. Properties are shown in Table 6.

Table 6. Properties of 100% polypropylene copolymer foam produced by the nitrogen autoclave process. Benefits include the higher temperature resistance and the recyclability of the foam

Property	Test method	Unit	Polypropylene foam (PA)
Density	ISO 845:1988	kg/m^3	30
Compression stress at 25% strain	ISO 3386/1:1986	kPa	60
Compression stress at 50% strain	ISO 3386/1:1986	kPa	150
Tensile strength	ISO 1798:1983	kPa	575
Elongation at break	ISO 1798:1983	%	24
Tear strength	ISO 8067:1991	N/m	875
Compression set	ISO 1856:C:1980	%	7

Table 7. Properties of conductive (CN) and static dissipative (SD) foam based on low density polyethylene produced using the nitrogen autoclave process

Property	Test method	Unit	LD 32 CN	LD 50 CN	LD 30 SD
Density	ISO 845	kg/m^3	30	50	30
Compression stress at 25% strain	ISO 3386/1	kPa	55	80	45
Compression stress at 50% strain	ISO 3386/1	kPa	130	175	120
Tensile strength	ISO 1798	kPa	370	610	300
Elongation at break	ISO 1798	%	50	50	65
Volume resistivity	ASTM D991	Ω.cms	5×10^3	1×10^3	-
Surface resistivity	EOS S-11.11	Ω/sq	-	-	1×10^7

SPECIALITY FOAMS

Incorporation of FR additives (antimony trioxide/halogen compounds) at the extrusion stage provides self-extinguishing foams; incorporation of electrically conducting additives, such as high structure carbon black, provides foams with electrical conductivity. A recent development has been foam having electrical conductivity in the static dissipative (SD) range (10^5-10^9 1/sq surface resistivity). SD properties are not easy to achieve by the use of conductive additives since conductivity is very sensitive to additive level in this range - foams tend to change from insulative (> 10^{12} 1/sq) to conductive (< 10^4 1/cm) with only a small increase in additive level. However, by suitable choice of carbon black it has been found possible to decrease this sensitivity, and this together with additional mixing control at the extrusion stage, has made possible the production of an LDPE foam with static dissipative properties, as shown in Table 7.

Table 8. Comparative properties of foams based on low density polyethylene with densities as low as 10 kg/m³, produced using the nitrogen autoclave process

Property	Test method	Unit	Foam density, kg/m³		
			10	24	45
Compression stress at 25% strain	ISO 3386/1	kPa	26	35	50
Compression stress at 50% strain	ISO 3386/1	kPa	86	100	135
Tensile strength	ISO 1798	kPa	174	340	600
Elongation at break	ISO 1798	%	100	105	150
Tear strength	ISO 8067	N/m	339	410	1030
Compression set	ISO 1856:C	%	12	8	7

FUTURE DEVELOPMENTS

In crosslinked foam processes other than the autoclave process it becomes increasingly difficult to manufacture foams having densities below about 25 kg/m³. The restrictions are partly technical (compounding problems with high levels of blowing agent, handling problems with very low density foam in the hot state and the presence in the foam of substantial amounts of solid residues) and partly economic (the high cost of additional blowing agent). The autoclave process allows lower foam densities to be produced by raising either the gassing temperature or the gassing pressure, both of which increase the quantity of nitrogen dissolving in the polymer, and hence the degree of expansion. Development materials with foam densities as low as 10 kg/m³ have been produced, and the physical limit of expansion has not yet been reached. Plant modifications are in hand to enable commercial production of 15 kg/m³ foam, initially based on LDPE, and properties of such materials are shown in Table 8.

ACKNOWLEDGMENTS

Thanks are due to the staff of the Technical department at Zotefoams plc for producing much of the data contained in this paper and to the management of the company for permission to publish.

REFERENCES

1   Eaves D.E., *Cellular Polymers*, **7**, 4, 1988, p 297.
2   Puri R.R. & Collington K.T., *Cellular Polymers*, **7**, 1, 1988, p 56.
3   Puri R.R. & Collington K.T., *Cellular Polymers*, **7**, 3, 1988, p 219.
4   Eaves D.E., 2nd Intl. Conference on Cellular Polymers, Edinburgh, 1993.
5   Oxley D., *European Plastics News*, May 1997, p 21.

Properties of Syndiotactic Polypropylene Fibers Produced from Melt Spinning

Mohan Gownder
Fina Oil and Chemical Company

INTRODUCTION

Research on sPP has gained momentum in recent years primarily due to its ease of synthesis. Even though Natta[2] reported the first preparation of sPP in 1960, it was not possible to pro-

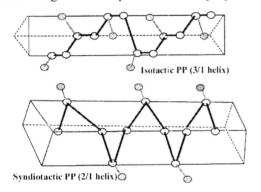

Isotactic PP (3/1 helix)

Syndiotactic PP (2/1 helix)

duce sPP of high stereo-regularity until recently.[3] Now, with the advent of metallocene catalysts, and coupled with their high activity, it is possible to synthesize propylene into a highly syndiospecific polypropylene.

The rules that govern the conformation of molecules, and their crystallographic packing (polymorphism) have been well established in the literature for sPP.[4] It's single chain conformation is helical with two methyl

(CH_3) groups per repeat unit (see picture). In comparison, there are three methyl groups per repeat unit for iPP.

The first crystal structure of a polymorph of sPP was reported by Corradini *et al.* in 1967 based on x-ray diffraction studies.[5] This is generally referred to as Cell-I in the literature and presented in Figure 1. Cell-I contains a C-centered structure with the helical chains packed isochirally (similar hand) in an orthorhombic unit cell with axes a=14.5 Å, b=5.6 Å and c=7.4 Å, where c is the chain axes. Recently, Lotz et al.[6] were successful in showing two additional interchain packing of the helical chains (shown as Cell-II and Cell-III in Figure 1.) based on their electron diffraction results. Cell-II has the same dimensions as Cell-I but the helical

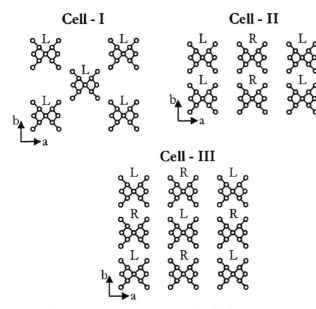

Figure 1. The three main polymorphic forms identified for sPP.

chains are placed in the ac faces of the unit cell rather than C-centered. Also, the molecules in the ac faces are antichiral, and in the bc faces they are isochiral. In Cell-III, the chains are packed antichirally in both a and b axes, and results in the doubling of the cell in the b axes (b=11.2 Å). Cell-III has been proposed as the most stable form of sPP and can be prepared at high crystallization temperatures, say about 140°C.[7]

Crystallization kinetics studies have shown that the molecular chain packing in sPP is very sensitive to its thermal history.[7-9] In Ref. 7, samples of sPP (>99% racemic dyad content), when isothermally crystallized at 140°C for an extended period of time, resulted in the fully antichiral stable Cell-III. Differential scanning calorimeter (DSC) thermogram showed a single peak when this sample was melted. In another study,[10] DSC thermogram showed two melt peaks possibly due to two different crystallographic packing patterns. As crystallization conditions during fiber spinning are extreme, it is critical to understand the nature of the packing of crystals during this process.

In most applications of polypropylenes, they are oriented either monoaxially (fibers, slit-tapes, etc.) or biaxially (films, sheets, etc.). Hence, it is important to study oriented polypropylenes and their mechanical behavior resulting from orientation. Sakata et al.[8] studied various sPP and iPP films, with and without orientation. They concluded, in the sPP samples studied, that factors such as drawing temperature, drawing speed, molecular weight and initial morphology show an insignificant impact on modulus. In sPP and iPP of comparable molecular weight, sPP always shows considerably lower modulus. To explain sPP's rubber-like behavior, they conclude that the absence of an α process, as observed through dynamic mechanical analysis, results in restricting the slippage of chains through the crystals in response to imposed macroscopic deformation. It is also hypothesized that the crystals behave as junction points.

In addition to several laboratory synthesis reported in the literature, the preparation of sPP has also been extended to commercial productions. Current attempts to commercialize sPP have aimed at the same applications isotactic polypropylenes are being used. In this study, a sPP and iPP of comparable MFI will be compared for their carpet fiber applications. It will be seen that even though sPP and iPP belong to the same group, as polyolefins, the properties vary widely. To study the structure, DSC and X-ray diffraction were used. All the fiber samples in this study were prepared in a melt spinning machine.

EXPERIMENTAL

Both the sPP and iPP used in this study were produced commercially in the pellet form by Fina Oil and Chemical Company in La Porte, TX. The sPP, synthesized using a metallocene catalyst, has a melt flow index of 8 and the iPP, synthesized using a Ziegler-Natta catalyst, has a melt flow index of 9. The molecular weight (M_w) for sPP is 107,000 and that for iPP is 248,000. The syndiotacticity of sPP measured by ^{13}C NMR shows the racemic pentad (rrrr) content to be 77% and the racemic dyad content to be 91%. The isotacticity measurement on the iPP shows the meso pentad content to be 91% and the meso dyad content to be 95%.

The melt temperature in a fiber spinning machine was set at 235°C for sPP and 230°C for iPP. Two 60-hole, trilobal spinnerettes were used to collect fibers with a linear density of about 1100 denier (denier = g/9000 m). One set of fibers, collected at various spinning speeds (hence orientation only in the melt state), was subjected to wide-angle x-ray diffraction (WAXD). Another set of fibers was collected at different draw ratios with the final speed kept at 1000 m/min. These were tested for their physical properties using an Instron tensile testing machine (elongation, tenacity, elastic recovery and permanent deformation when cyclically deformed). For the elastic recovery test, a deformation sequence based on ASTM D1774-90 is followed. It consists of a constant rate (10%/min) loading and unloading, with a maximum strain of 10% (and also 50%) maintained for 1 min duration in between the ramps. This cycle is repeated after a 3 min interval. Fibers for both sPP and iPP drawn at 3:1 were also tested for elongation and tenacity after 3 months to determine if aging had any effect on the properties.

WAXD patterns were recorded at room temperature with a Siemens diffractometer using Ni-filtered CuK$_\alpha$ radiation (λ= 1.54 Å) and a step-scan procedure. The range of 2θ diffraction angle examined was between 3 and 35° with a step-scanning rate of 0.08°/sec and a counting time of 8 sec at each step. For the fiber samples, x-ray diffraction was along the fiber axis (meridional scans).

Dynamic mechanical tests were performed on the fibers using a rheometer (Mark-IV of Rheometrics Scientific). The fibers were deformed sinusoidally in the axial direction with a static deformation also applied on the fiber during the test. The tests were conducted in the linear viscoelastic range. Dynamic moduli (G' and G'') and tan δ were measured for a temperature range of -30 to 100°C at 10 Hz.

RESULTS AND DISCUSSION

Table 1. Processing conditions during fiber spinning

Property	iPP	sPP
Melt temperature	230	235
Spin tension at 3:1 draw and G2= 1000 m/min (gf)	54	-
Draw tension at 3:1 draw and G2=1000 m/min (gf)	2000	150
Max draw at G2=1000 m/min	3.6	3.6

During melt spinning, the polymer melt is formed into filaments by extrusion through a spinnerette. After the filaments thin down and solidify, they are also oriented (drawn) to improve their mechanical properties. Table 1 shows the processing conditions during the spin and draw of sPP and iPP fibers. The sPP fibers extruded from the spinnerette were sticky due to their low solidification rate. Hence, it was not possible to measure its spin tension (the strands stuck to the tensiometer). The very low draw tension of sPP was due to its extreme softness even until the drawing stage. Increasing the quench rate or reducing the quench temperature did not improve this condition. The maximum draw ratios for both the materials were the same. The sPP fibers produced were also elastic in nature and texturizing the fibers did not provide enough bulk to the yarn. The fibers were soft compared to a commonly used bulk-continuous-fiber (bcf).

As has been observed in the literature, the as-prepared sPP tested in the DSC showed two melting peaks (T_m), one at 115°C and the other at 130°C. The iPP melting peak was at a much higher temperature of 160°C. Study of fiber structure dependency with orientation using DSC will improve our understanding of the melt spinning of sPP. This will be the subject of a future study.

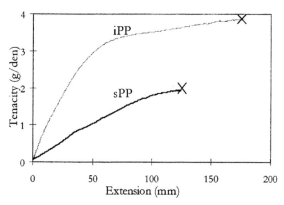

Figure 2. Tenacity vs. extension for both sPP and iPP fibers drawn at 3:1 during fiber spinning.

Figure 2 shows the tenacity of sPP and iPP fibers, drawn to 3:1 ratio during melt spinning, extended until break. Figures 3 and 4 show the maximum elongation and tenacity of the fibers prepared at various draw ratios. The elongation for sPP fibers is independent of the draw ratio. Normally for an iPP, as seen in Figure 3, the elongation decreases with draw ratio. In iPP fibers, an increase in melt orientation aligns the molecular chains in the direction of the fiber axis and upon solidification the molecules are re-

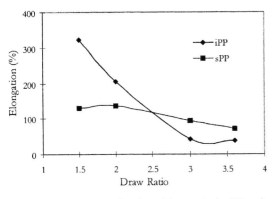

Figure 3. Elongation as a function of draw ratio for iPP and sPP fibers.

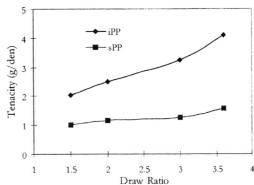

Figure 4. Tenacity as a function of draw ratio for iPP and sPP fibers.

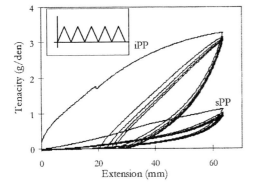

Figure 5. Tenacity vs. extension for sPP and iPP fibers (3:1 draw) for 5 cycles. The cyclic input is shown in the inset.

tained in the same state. In the case of sPP fibers, since solidification is delayed, the molecules have time to relax and keep the same conformation even though they are extended in varying degrees during the melt spinning process. This can also be seen in Figure 4 for the plot of tenacity vs. draw ratio. The tenacity for sPP fibers is independent of the draw ratio whereas for iPP, it increases with draw ratio. It has been surmised in the literature[11] that the presence of entanglements and crystallites in sPP prevent molecules from sliding past each other thereby reducing the maximum elongation.

The free shrinkage of the two fibers were tested at 125°C. The sPP shows a much higher shrinkage compared to iPP. This is probably due to the highly entropic nature of drawn sPP, and when the thermal energy is supplied, the stretched molecular chains revert to their original conformation thus causing excessive shrinkage in the fiber.

During processing in a textile operation, the fibers come under varying degrees of load. To estimate the capability to withstand these loads and also to estimate the elastic recovery without any permanent deformation, the two sPP and iPP fibers were subjected to load-deformation and cyclic tests. Figure 5 shows the tenacity at 5 consecutive cycles of loading and unloading of the sPP and iPP fibers. The elastic nature of sPP is quite evident in that

Table 2. Elastic recovery measurements on sPP and iPP

Material	% max def.	iPP	sPP
Permanent deformation	10	6	0
	50	43	0
Work recovery	10	35	55
	50	16	21

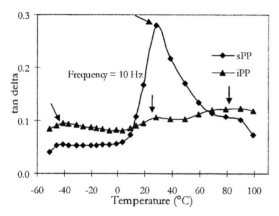

Figure 6. Tanδ as a function of temperature at 10 Hz. The arrows point the relaxations that are taking place.

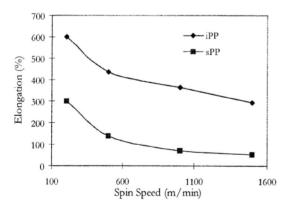

Figure 7. Elongation of sPP and iPP fibers partially oriented at various spinning speeds.

the permanent deformation is nonexistent whereas iPP suffers considerable permanent deformation.

Table 2 shows the results of ASTM testing conducted on these samples which evaluate the permanent deformation and work recovery. Even with time allowed for recovery, the iPP shows a high permanent deformation.

Results of dynamic mechanical tests performed on these fibers are shown in Figure 6. As observed by Sakata et al.,[11] for stretched films, sPP fibers showed only one relaxation whereas iPP fibers showed two relaxations (noted by arrows).

Polypropylenes, both iPP and sPP, have been known to exhibit changing properties with time. It is generally accepted that polypropylenes do not crystallize completely when cooled from the melt to the solid state and that crystallization continues for an extended period of time. Using these fibers, changes in properties with the slow crystallization phenomena of sPP were evaluated by testing the specimen collected immediately after spinning, and the same specimen about 3 months after collection. In general, both sPP and iPP showed a decrease in elongation and tenacity with time. This is certainly due to the slow crystallization taking place in the polymers

Oriented sPP studied in the literature so far has been that of quenched plates or films oriented by stretching. In this study, the samples oriented during fiber spinning in the melt state is discussed. Figures 7 and 8 show the elongation and tenacity of the

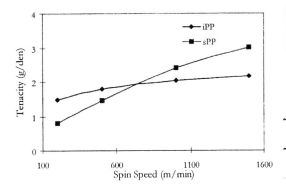

Figure 8. Tenacity of sPP and iPP fibers partially oriented at various spinning speeds.

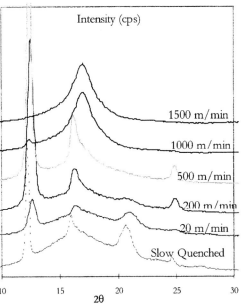

Figure 9. WAXD patterns for sPP collected at various orientations.

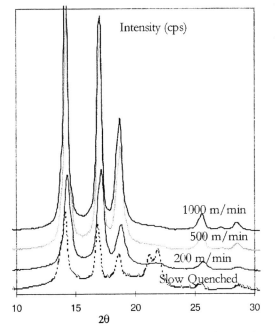

Figure 10. WAXD patterns for iPP collected at various orientations.

fibers spun at various spinning speeds without any drawing. The structures of these fibers as observed through I-2θ plots are presented in Figures 9 (for sPP fibers) and 10 (for iPP fibers). It has been shown that the structure and morphology of sPP are very sensitive to the temperature of crystallization.[9] Lovinger *et al.*[1] have shown that when sPP was melt grown (crystallized) at higher temperatures, the crystallographic structure formed was without defects. Lowering the temperatures (<80°C) progressively introduced packing defects as shown by streakings in the electron diffraction patterns (kinetic factors favor this). With regard to packing, the placement of the two hands of the helices is what determines crystallographic packing. The disorder was explained as the incorporation of chains in the c-centered (Cell I) manner together with row vacancies.[6] Lovinger *et al.*[1] concluded that in single crystals of highly syndiotactic polypropylenes, the preferred mode of packing is Cell III. In Figure 9, the occurrence of 010 reflection at $2\theta=16.0°$ reveals the packing mode of the kind Cell II. Fiber samples collected with slight orientation, at 20 m/min, the three basic reflections are present. With increasing speed (as a result melt orientation), the 100 peak gets sharper. The 010 peak gets higher in intensity and at the same time, broader. When the speed is increased from 500 m/min to 1000 m/min, there is a drastic change in the fiber spectra as 100 reflection almost disappears and the 010 peak widens further. At 1500 m/min, there can be seen only one broad peak. According to Lovinger,[1] the breadth of the peak is a result of substantial packing disorder. Figure 10 shows the reflections of the iPP fibers collected at various spinning speeds. The crystalline content of iPP increases with spin speed as the peaks are getting sharper with increase in speed. The x-ray diffraction patterns clearly differentiates the crystallization tendencies of the sPP and iPP fibers.

Lovinger *et al.*[1] also identified anisotropic thermal properties and the mobility of point vacancies and state these have a profound effect on the weakening of the mechanical properties of sPP. This is severe in processes such as fiber spinning where rapid cooling of the sPP melt is involved.

CONCLUSIONS

The crystallographic packing of the sPP molecules are highly dependent on the thermal and mechanical history. It is shown here, from x-ray diffraction patterns, that quenching the sPP melt at lower temperatures (due to cool air at 10°C from the quench column of fiber spinning machine) and simultaneously orienting has introduced packing defects in the structure of sPP fiber. It was proposed earlier[1] that this would have a negative effect on the mechanical properties of sPP. This is reflected in the lower elongation and tenacity of the spun and drawn fibers. However sPP possesses a high work recovery and low permanent deformation which are essential for spinning carpet fibers.

ACKNOWLEDGMENTS

The author gratefully acknowledges the assistance of Jay Nguyen in preparing the fiber samples, Ed Zamora, Sam Curtis and Rebecca Davis in characterization. The author wishes to thank Al Baumgartner and David Rauscher for their helpful conversations during this study. The author also wishes to thank Michel Daumerie, and Fina Oil and Chemical Company for their encouragement and support.

REFERENCES

1   Lovinger, A.J.; Lotz, B.; Davis, D.D.; Schumacher, M., *Macromolecules*, 1994, **27**, 6603.

2   Natta, G.; Pasquon, I.; Corradini, P.; Peraldo, M.; Pegoraro, M.; Zambelli, A., *Rend. Acd Naz. Lincei*, 1960, **28**, 539.
3   Ewen, J.A.; Jones, R.L.; Razavi, A.; Ferrera, J.D., *J. Am. Chem. Soc.*, 1988, **110**, 6255.
4   Lotz, B.; Wittmann, J.C.; Lovinger, A.J., *Polymer*, 1996, **37**, 4979.
5   Corradini, P.; Natta, G.; Ganis, P.; Temussi, P.A., *J. Polym. Sci., Part C*, 1967, **16**, 2477.
6   Lotz, B.; Lovinger, A.J.; Cais, R.E., *Macromolecules*, 1988, **21**, 2375.
7   Lovinger, A.J.; Lotz, B.; Davis, D.D.; Padden, Jr. F.J., *Macromolecules*, 1993, **26**, 3494.
8   Lovinger, A.J.; Davis, D.D., *Macromolecules*, 1991, **24**, 552.
9   Rodriguez-Amold, J.; Zhang, A.; Cheng, S. Z. D. et al., *Polymer*, 1994, **35**, 1884.
10   Loos, J.; Buhk, M.; Petermann, J.; Zoumis, K.; Kaminsky, W., *Polymer*, 1996, **37**, 387.
11   Sakata, Y.; Unwin, A.P.; Ward, I. M., *J. Mater. Sci.*, 1995, **30**, 5841.

Melt Spinning of Syndiotactic Polypropylene: Structure Development and Implications for Oriented Crystallization

Ravi K. Sura, Prashant Desai and A.S. Abhiraman
Polymer Education & Research Center, Georgia Institute of Technology Atlanta, GA 30332

INTRODUCTION

Stereoregular polymers prepared using metallocene catalyst systems have become quite important. The different molecular configurations of these polymers leads to changes in crystallization and melting behavior. These changes affect the properties of products made from these polymers. In the present study, syndiotactic polypropylene (sPP) was melt spun and the fibers analyzed to understand development of structure and morphology.

Most studies of sPP have focused on conformational structure and the nature of crystals, usually in the form of isotropic films. Only a few studies[1,2] have examined oriented products such as fibers. It is well known that orientation has a profound effect on the rate and extent of crystallization in melt spinning. A theoretical framework has been developed that explains the enhancement in crystallization kinetics because of enhanced orientation.[3] Higher strain rate, at higher spinning speed, results in greater segmental orientation prior to crystallization. However, it can also change the conformational composition of the sPP segments and, consequently, the crystallization behavior of oriented sPP filaments.

EXPERIMENTAL DETAILS

MATERIALS AND PROCESSING

Syndiotactic polypropylene (sPP) with melt flow rate of 8 at $230^{\circ}C$ was used in this work. Melt spinning was conducted at $230^{\circ}C$ in a PP research melt spinning line at a commercial research laboratory. A 7-hole spinneret was used, fed from a twin-screw extruder via a metering pump. Take-up speeds were varied from 100 to 2000 m/min with the throughput adjusted to give a linear density of 2 denier per filament (denier = g/9000 m).

CHARACTERIZATION TECHNIQUES

The melt spun fibers were studied using WAXD, thermal analysis (DSC and DMS), tensile testing, sonic modulus, birefringence, and solid-state CP/MAS measurements.

Since fibers are usually drawn to obtain superior properties, a simulated drawing experiment was also conducted by Thermal Deformation Analysis (TDA). This involves the measurement of changes in length of filaments when they are subjected to a constant axial force and heated rapidly to a temperature between T_g and T_m. The apparatus consists of a temperature controlled tubular heater mounted on a mobile platform that can be moved rapidly over rails to enclose the loop of filaments being tested. The fiber is connected at each end to Kevlar yarns, one of which is connected to a fixed support while the other is passed over a smooth pulley and attached to a weight. A pointer is attached to the moving end so that the deformation can be read against a scale. The deformation of the Kevlar yarn can be neglected when compared to that of a filament loop. In one type of this analysis, called the Isothermal Deformation Analysis (ITDA), the temperature of the heater is set and the experiment is started by hanging a constant weight to the fiber loop. The heater is then moved rapidly over the fiber so that it completely encloses it, maintained in this position for a period of 10 seconds to ensure complete deformation of the fiber under the applied stress and temperature, and rapidly withdrawn. The deformation is obtained from the change in length of the loop. The fiber samples after deformation are collected for analysis.

A similar set-up was used to perform a non-isothermal stress analysis (NTSA) where the fiber is attached to a load cell at one end and constrained at the other end. It is enclosed in a tubular heater and the temperature increased from room temperature to the failure point of the fibers. The stress developed in the fiber, as a function of temperature, is recorded.

RESULTS AND DISCUSSION

sPP was melt spun at various wind-up speeds ranging from 100 to 2000 m/min. It was observed that at spinning speeds greater than 1000 m/min, the fibers tended to stick to each other on the take-up roll, which was not the case for lower speeds. This suggested that the high speed spun (HSS, greater than 1000 m/min) fibers were not as well crystallized. Flat plate WAXD photographs (Figure 1) show that the low speed spun (LSS, up to 1000 m/min) fibers are distinctly semi-crystalline with high crystal orientation. The WAXD data also suggest that there is a change in the crystal structure for the high speed spun.

DSC scans for the various fibers are shown in Figure 2. The typically observed multiple melting peaks for sPP can be seen for the LSS fibers, while the higher melting peak disappears for the HSS fibers. It has been suggested in the literature that these peaks arise due to melting and recrystallization while heating the sample. However, there was no shift in the

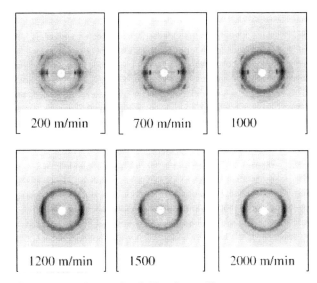

Figure 1. X-ray photographs of sPP melt spun fibers.

melting peaks when the heating rate was increased to 40°C/min, which suggests that the peaks arise from factors other than purely kinetic ones. Cold crystallization is detected in HSS fibers around 75°C.

Sonic modulus and birefringence decreased with spinning speed (Figure 3), indicating a decrease in overall orientation with increase in spinning speed. However, the extent of free shrinkage at 100°C increases from 10% to 70% as the spinning speed increases from 200 to 2000 m/min (Figure 4). These data are consistent with increased orientation of chains in the spinline at higher speeds, but higher extent of oriented crystallization at the lower speeds. The tensile properties, shown in Figure 5, support this view. The breaking stress increases and elongation to break decreases with increasing speed.

Non-isothermal stress analysis results are shown in Figure 6. Initially there is a reduction in the shrinkage force due to thermal expansion and orientational relaxation. The LSS fibers show an increase in shrinkage force close to the melting point (~130°C), possibly due to progressive melting of oriented crystals and the tendency of the oriented noncrystalline chain

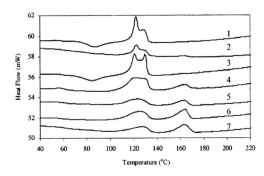

Figure 2. DSC scans of sPP melt spun fibers: 1 - 2000, 2 - 1500, 3 - 1200, 4 - 1000, 5 - 700, 6 - 400, 7 - 200 m/min.

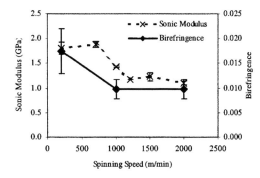

Figure 3. Sonic modulus and birefringence of sPP melt spun fibers as a function of spinning speed.

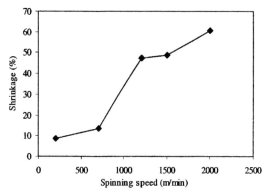

Figure 4. Shrinkage of sPP melt spun fibers as a function of spinning speed (3 min @ 100°C.)

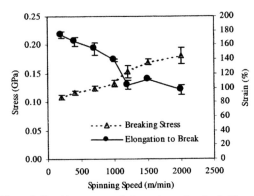

Figure 5. Breaking stress and elongation to break of sPP melt spun fibers.

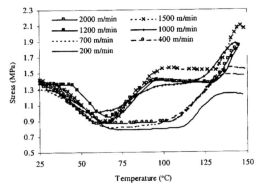

Figure 6. Non-isothermal stress analysis of sPP melt spun fibers - stress development as a function of temperature in a constrained fiber.

segments thus formed to coil. However, the HSS fibers show an increase in shrinkage force immediately after the relaxation phase is over. This could be due to chains in the all *trans* conformation converting to the helical -$(T_2G_2)_2$- conformation, a state in which they can crystallize. Calculations show that the conformational change requires approximately 40% reduction in length. Close to the melting point, the HSS fibers behave in a manner similar to LSS fibers.

In order to examine the conformational sequences in the sPP fibers, solid-state CP/MAS NMR spectroscopy was conducted. Melt-cast and solution-cast samples showed the stable, preferred conformation of a -$(T_2G_2)_2$-helical structure. The LSS fibers also exhibit a -(T_2G_2)- conformation, albeit in a different crystal structure as seen from WAXD equatorial scans. This gradually transforms to an all *trans* -(TTTT)- conformation in HSS fibers. The peak assignments were based on literature values.[4-7]

NMR measurements were also taken on fibers that were annealed under constraint at 100°C for 3 minutes in an air-circulated oven, to simulate the conditions seen in nonisothermal stress analysis. It was seen that the LSS fibers continue to remain in the helical -(T_2G_2)-conformation, while the HSS fibers transform from an all *trans* to a helical -$(T_2G_2)_2$- conformation.

The as-spun fibers were subjected to nonisothermal deformation analysis to observe the change in structure under simulated drawing conditions. At low stress levels, the fibers exhibit shrinkage, while higher stress levels are required to "draw" the fibers. The LSS fibers undergo lower shrinkage at low stress levels and higher elongation at high stress levels, when compared to the HSS fibers, consistent with the lower noncrystalline orientation in the LSS fibers. The noncrystalline chains in the LSS fibers can be extended from their almost helical conformation to a more stretched conformation. X-ray scattering reveals that the fibers subjected to nonisothermal deformation analysis are well crystallized and oriented.

The sPP melt spun fibers were subjected to constrained length annealing (CLA) and free length annealing (FLA). WAXD data suggest that while LSS fibers do not significantly change their crystal structure, the HSS fibers become distinctly semi-crystalline and attain a crystal structure, similar to the LSS fibers, on annealing. The higher sonic modulus on CLA is consistent with the development of oriented amorphous phase in CLA fibers.

It is well known that the most stable and preferred crystal structure conformation for sPP is that of helical $-(T_2G_2)_2-$ sequences. At low spinning speeds, the chains do not get stretched too much out of their helical conformation. As the polymer fiber cools down, crystals are easily formed as the chains are close to the preferred conformation. However, as the spinning speed increases, the stress increases and the chains are stretched into the extended all *trans* conformation, which is not favorable for crystallization. Thus, there is a decrease in the rate of crystallization. Coupled with the lesser time for crystallization at higher speeds, this results in reduced extent of crystallization. On annealing HSS fibers, even for relatively short times, the chains revert to the preferred helical conformation. This happens even under constrained conditions, as there are not many physical constraints against conformational changes due to the absence of crystals. The appropriate conformational sequence can have a rapid rate of incorporation into crystals, due to high chain orientation.

However, on FLA, the HSS fibers attain a relaxed amorphous phase and a stable helical crystal structure. The relaxed amorphous phase contributes to large extensions ($\sim$450%) observed for HSS fibers which were subjected to FLA.

Our results show that the sPP fibers are demonstrably poorly crystallized at higher spinning speeds in contrast to well crystallized, oriented fibers formed at lower spinning speeds. This behavior occurs at a relatively low spinning speed. The HSS fibers have the more extended $-(TTTT)-$ conformation compared to the helical $-(T_2G_2)-$ conformation in the LSS fibers. However, crystallization of oriented all *trans* sequences would require change in conformation to form helical sequences that can crystallize. Thus the rate of crystallization is, in fact, reduced. This effect must be taken into account in developing theories of oriented crystallization. Work is underway to address this aspect of oriented crystallization.

ACKNOWLEDGMENTS

The authors express their gratitude to Dr. Mohan Gownder and Dr. Joseph Sehardl of Fina Oil and Chemical Company for supplying the polymer. They also thank Karthik Nagapudi for the NMR experiments.

REFERENCES

1   Gownder, M., ANTEC Conf. Proc., 1998,1511.
2   Chatani, Y.; Maruyama, M.; Asanuma, T.; Shiomura, T., *J. Polym. Sci., Pt. B: Polym. Phys.,* 1991, **29**, 1649.
3   Desai, P.; Abhiraman, A. S., *J. Polym. Sci., Part B: Polym. Phys. Ed.*, 1989, **27**, 2469.
4   Asakura, T.; Aoki, A.; Date, T.; Demura, M.; Asanuma, T., *Polym. J.*, 1996, **28**(1), 24.
5   Auriemma, F.; Born, R.; Spiess, H.W.; Rosa, C.; Corradini, P., *Macromolecules*, 1995, **28**, 6902.
6   Sozzani, P.; Monutti, R.; Galimberti, M., *Macromolecules*, 1993, **26**, 5782.
7   Sozzani, P.; Galimberti, M.; Balbontin, G., *Makromol. Chem., Rapid Commun.,* 1992, **13**, 305.

New Polyolefins Characterization by Instrumental Analysis

Carmina Gartner and Juan D. Sierra
*Instituto de Capacitación e Investigación del Plástico y del Caucho, ICIPC, Medellin,
Colombia, P.O. Box 2629*
Roger Avakian
General Electric Specialty Chemicals, Parkersburg, WV, USA

INTRODUCTION

Given the multiple synthesis methods, polyolefins, PO, are found within a wide property range that are determined primarily by their basic chemical structure. The study of their chemical features can be difficult due to the many possibilities of structural arrangements and increasing complexity which requires the conjunction of many instrumental techniques.

The main purpose of this research is to define the structure of PO, basically polyethylene, PE and polypropylene, PP of highest commercial importance, through a comparative study of the infrared spectra, in order to find all the information which is possibly available by this technique. Besides infrared spectroscopy, differential scanning calorimeter results, DSC, are frequently used for determining certain important thermal properties directly associated with morphology and crystallinity.

EXPERIMENTAL

This report data analysis is based on measurements made on Fourier Transform Infrared Spectrophotometer, FTIR, Nicolet Magna-IR 550, with resolution of 0.50 cm^{-1}, autoaligning digital system and detector in the middle infrared range, and a Differential Scanning Calorimeter, DSC, TA Instruments, 2910, a heat flow type calorimeter, with an standard cell.

The tested materials were commercially used PO, except for three of the PP, which were kindly supplied by the manufacturer from the polymerization reactor ; none of them had additives that could affect the results. The analyzed PE, were chosen to be from different

Table 1. Characteristic IR absorptions for PE: wavenumbers between 1500 and 700 cm^{-1}

λ, cm^{-1}	Assignment	λ, cm^{-1}	Assignment
1471	symmetric bending of methylene (tentative)	910	terminal groups along with 990
1463	asymmetric bending of methylene	908	also associated with terminal vinyl groups
1378	symmetric bending of methyl	903	along with 889 it is assigned to terminal methyl groups from alkidic chains longer than ethyl
1367	scissoring of methylene	889	bending vibrations of methyl groups from n-hexyl. If there is an overtone in 745 it corresponds to n-butyl
1353	torsion bending of methylene	888	vinylidene type unsaturations
1346	bending of methylene for a regular pack structure	849	vibration of CH$_2$ related with a crystalline conformation of the chain
1304	bending of methylene associated with the amorphous part	839	possibly due to 3 substituted alkenes, like a diallyl group (tentative)
1176	wagging of C-C from methylene	802	like the one at 839 (tentative)
1151	vibration of C-C from methyl	784	vibrations of methylene longer than ethyl (tentative)
990	vibration of CH of terminal vinyl groups	762-70	vibrations of CH from ethyl group (two consecutive CH$_2$ groups)
967	Vibration CH of vinylidene groups (internal trans unsaturations)	745	this band along with the one at 890 can be related to n-butyl branches
954	vibration of the allyl type (tentative)	730	indicative of the long aliphatic semicrystalline chains of carbon bonds. Attributed to the rocking of methylene in the crystal of polyethylene
937	out of plane tension of terminal methyl groups	720	rocking vibrations of minimum 4 consecutive CH$_2$ groups

The assignations for the bands in 1471, 954, 859, 839, 802 and 784 cm^{-1} are tentative and have not been well correlated for a specific vibration.

polymerization processes and densities: high pressure, Ziegler and Phillips processes and metallocene catalysis, varying from low to high density. PP were homopolymers and ethylene copolymers.

For FTIR measurements, the tested materials were prepared as thick films, by heating the samples between two metallic plates inside an aluminum foil, at approximately 200°C for half an hour, then quench cooled to avoid light scattering due to crystals, measuring the thickness with a Vernier and taking the spectra by the transmission technique at the same

Table 2. Characteristic IR absorptions for polypropylene: wavenumbers between 1500 and 700 cm^{-1}

λ, cm^{-1}	Assignment	λ, cm^{-1}	Assignment
1467	scissoring bending of methylene	972	*assigned to two or more head to tail units in PP
1456	asymmetric bending of methyl	960	sequence of two contiguous units in polypropylene
1379	symmetric bending of methyl	940	rocking of methyl group
1356	scissoring bending of methylene	899	*assigned to a helicoidal conformation of polypropylene
1304	torsion and wagging stretching of methylene	840	*helicoidal conformation. It is also attributable to a length of isotactic sequences of 13 to 15 units
1255	vibrations of carbon backbone for C-CH$_3$ bonding	809	*related also to helicoidal conformation similar the other bands with asterisks at 1219, 1167, 997, 899 and 840
1219	*attributed to the helicoidal conformation of polypropylene	732	only presented in random copolymer, due to a sequence of 3 contiguous CH$_2$ groups
1167	*as the band at 1219	728	sequence of four contiguous methylene groups
1153	stretching wagging of methyl assigned to solely propylene units, presented on in copolymers	720	only presented in block copolymer. Assigned to a sequence of five or more contiguous methylene groups
997	*helicoidal conformation. It also can be due to a length of isotactic sequences from 11 to 12 units		

Bands with asterisks (*) correspond to the helicoidal conformation of polypropylene associated with an isotactic structure

conditions. For DSC measurements, the materials were cut until obtaining the appropriate weight of 10 ± 1 mg, then samples were heated and annealed at 200°C for 10 minutes, (to erase the thermal history), next they were cooled at a rate of 10°C/min and reheated at the same rate to study the melting curve.

RESULTS AND DISCUSSION

INFRARED ANALYSIS, IR

The analyzed IR spectra holds the region between 1500 and 400 cm^{-1} as wavenumbers, λ, which has been studied on what refers to PO microstructure. All the absorption bands were

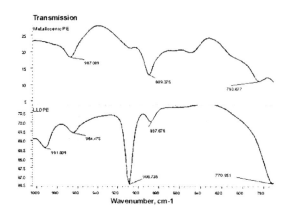

Figure 1. IR spectra of PP.

tabulated and assignments were found for all particular molecular vibrations which are summarized in Tables 1 and 2.[2,3,4,5,6,9]

In the region between 1500 and 1300 cm^{-1} of CH bending, the main difference lies in the intensity of the band associated with methyl groups in 1378 cm^{-1}; its presence is an indication of long branching content that can be used for means of quantification.[4] The region between 1300 and 1000 cm^{-1}, carbon backbone stretching, is associated with crystallinity, and deserves more deep studies. The region between 1000 and 700 cm^{-1} is commonly used for identification and quantification of branching, unsaturations and crystallinity. Below 720 cm^{-1} there were no bands of analytical significance. Next are described the principal qualitative characteristics encountered for the different tested materials.

In high density, HDPE, the unsaturations of terminal vinyl groups predominate because of the presence of the bands in 990 and 908 cm^{-1} and because there is no absorption at 888 cm^{-1}, it can be concluded that there are no unsaturations of the vinylidene type groups. This is common in a PE produced by the Ziegler process in which 43% of the unsaturations are of the vinyl type.[11] The absence of the methyl group in plane vibration at 1378 cm^{-1} is indicative of a low long branching content. The spectrum for medium density, MDPE, is very similar to the one of HDPE, although there is an increase of the vinylene type unsaturations which are seen in the intense band at 965 cm^{-1} (internal *trans* unsaturations). The predominant unsaturations are of terminal vinyl type (909 and 990 cm^{-1}), characteristic of a Phillips PE.

For linear low density, LLDPE, because of the presence of a very intense band at 908 cm^{-1} from vinyl and the increase of the vinylidene type unsaturations, it was seen that this material corresponds with a Ziegler PE.[11] It is observed a characteristic band at 770 cm^{-1} that corresponds to ethyl groups from copolymerization with 1-butene (see Figure 1).

For low density, LDPE, the unsaturations of the vinylidene type are predominant over the vinyl ones, which can be seen from the very intense band at 888 cm^{-1} what could be expected from high pressure PE, where vinylidenes are the 68% and vinyls are the 15% of all the unsaturations. It was observed that unsaturations of the vinylidene type are more frequent in highly branched PE or those with lower molecular weight. The absorption at 957 cm^{-1} may be due to allyl type unsaturations.

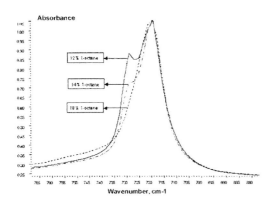

Figure 2. Effect of concentration of 1-octane on IR.

For metallocenes PE the intensity of the bands corresponding to unsaturated groups are much lower. The band in 888 cm^{-1} moves to 889 cm^{-1} where the vibrations of saturated chains greater than two carbons are presented and that could correspond to hexyl groups originated from the comonomer 1-octene; even though the resolution of 0.5 cm^{-1} would be enough to resolve these two bands, it becomes necessary to eliminate possible unsaturations to correct overlapping. It is seen from absorption at 784 cm^{-1} which can be related to longer branches than ethyl, (probably hexyl) (see Figure 1).

For ethylene vinyl acetate copolymers, EVA, the vibrations of ester groups predominate and the basic differences between these copolymers are the intensity of the carbonyl tension, which increases along with the comonomer content and that is used for quantification. There are ethyl branches and unsaturations of the vinylidene, allyl and vinyl type, characteristic of a highly branched PE (produced by the high pressure method).

The presence and intensity of the two bands at 730 cm^{-1} and 720 cm^{-1} are indicators of long aliphatic semi-crystalline chains. The intensity of the band at 730 cm^{-1} increases with crystallinity and hence with density of PE, and diminishes for the lower crystalline materials. For metallocenes it can be seen the decrease in absorption for the band at 730 cm^{-1} as the comonomer content increases, because the addition of a comonomer disrupts the stereoregularity and thus the crystallinity (see Figure 2) - a feature that can be observed for LLDPE and EVA copolymers, as well.

Polypropylenes: For PP two regions were studied: the CH bending between 1500 and 840 cm^{-1} (corresponding to tensions of the carbon backbone chain), that presented a sequence of bands that deal with stereoregularity, and between 840 and 400 cm^{-1} (assigned to CH out of plane bending), where the vibrations for comonomer sequences are seen.

For all the analyzed PP, the characteristic absorptions of isotactic segments at 1219, 1167, 997, 899, 840 and 809 cm^{-1} were seen. Particularly the bands at 997 and 840 cm^{-1} show that the isotactic segments are long, from 11 to 15 units, and their intensities ratio can be used for measuring the degree of isotacticity. It can also be observed that PP are highly symmetrical presenting many head to tail bonds, because of the intense band at 972 cm^{-1}. For none of the PP it was possible to see the atacticity or syndiotacticity fractions that show characteristic

Table 3. Results of calorimetric measurements

Polyolefin type	Sample identification	Melting temperature, °C		Heat of fusion, J/g
		T_1	T_2	
LDPE	L1	112.19	-	99.05
	L2	112.4		106.5
LLDPE	LL1	108.88	123.95	119.3
	LL2	108.49	121.15	110.5
	LL3	107.65	123.37	129.2
	LL4	110.64	123.52	108.8
	LL5	not resolved	125.93	101.01
MDPE	M1	125.02	-	153.4
HDPE	H1	132.27	-	208.7
	H2	135.16		224.7
	H3	132.45		186.6
	H4 (bimodal distribution)	131.06		193.5
Metallocenes	VL1 (12% 1-octene)	98.92	-	63.15
	VL2 (14% 1-octene)	94.31		60.83
	VL3 (18% 1-octene)	67.13		23.56
EVA	EVA1 (9% vinyl acetate)	97.35	-	66.12
	EVA2 (18% vinyl acetate)	85.62		58.77
PP	H1, homopolymer	166.39	-	99.80
	H2, homopolymer	162.96	-	97.32
	H3, ethylene block copolymer	115.91	163.66	85.32
	H4, ethylene block copolymer	141.71	-	80.22

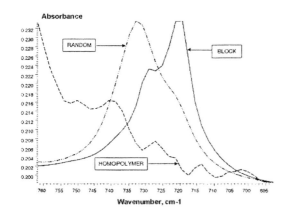

Figure 3. Comparison of different polymers.

bands at 815 and 752 cm[-1] with sequences of two or more contiguous methylene groups and definite quantities of head to head and tail to tail unions.[5] All the above reasoning demonstrated the high stereoregularity of these materials.

Impact or block copolymers presented bands at 729 and 720 cm[-1] corresponding to long sequences of methylene groups, while the random type presents a band at 730 cm[-1] attributed to a sequence of three consecutive

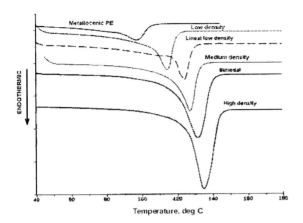

Figure 4. Melting behavior.

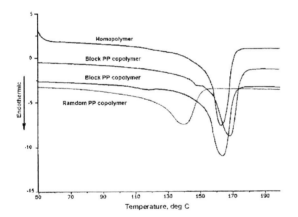

Figure 5. Melting behavior.

methylene groups and does not show a band at 720 cm^{-1} (see Figure 3).

DSC RESULTS

The heat of fusion reported corresponds to the area under the curve and the melting temperature is the peak maximum, calculated by a software. The calorimetric results for the analyzed samples are summarized in Table 3.[8,12]

Polyethylenes: As it was expected, the lowest melting temperatures and enthalpies are associated with a less crystalline structure and a high branching content. It had been demonstrated that density of PE is directly associated with crystallinity,[6] and when it was plotted against melting temperature for several tested PE, there was a linear variation and a high correlation coefficient of 0.9774. It is also observed that the lower the degree of crystallinity the wider the melting curve, behavior that could be explained by the large distribution of crystal sizes, crystalline imperfections and lower molecular weight (see Figure 4). LLDPE presents a higher melting point in comparison with LDPE, and shows two melting point distributions (a maximum around 125°C and a small one around 110°C); this can be explained by the assumption of the presence of two different crystallite size forms in the molecule attributable to a different ethylene sequence length distribution, which could be the result of many variables during the synthesis process.[1,7,8] This behavior is not observed for the homopolymer MDPE of approximately the same density as LLDPE, although their have similar melting temperature around 125°C.

Metallocenes PE, although copolymers, did not present the molecular hetereogeneity explained above for LLDPE, due to the high stereospecificity of these catalysts and presented the lowest melting points due to their high comonomer content and their low molecular weights. It can be seen from data in Table 3 that with the increase of comonomer content, the temperature and heat of melting decreases. The same reasoning can be made for EVA copolymers when the vinyl acetate content is increased.

Polypropylenes: The heat of fusion and melting temperature for PP homopolymer is higher in comparison with its ethylene copolymers (see Table 3 for results). The introduction of an ethylene comonomer changes the thermal behavior of PP. In a random distributed addition of comonomer, the copolymer forms mixed crystals sharing a common crystal lattice (isomorphism), or else only one of the comonomer unit crystallizes and the other remains in the amorphous region.[5,10] which explains the lower crystallinity when compares with the homopolymer. For the block copolymer there is a sequential distribution of comonomer which permits each fraction of them to crystallize in their own structure as in the corresponding homopolymer; this is seen in the two melting peaks obtained for the block copolymer: 115 and 163°C (see Figure 5).

CONCLUSIONS

PE of different densities can be distinguished by certain characteristic features in the structure which are perfectly seen in the IR spectra and in the DSC curves. These differences are mainly short branching as hexyl and ethyl, unsaturations of the internal *trans* vinylene, vinyl, vinylidene and allyl type. The typical absorption bands that were found may be used for future quantitative studies of degree of short and long chain branching, unsaturations and percentage of comonomer content. For the PP tested the series of IR bands related to isotacticity were detected which opens the possibility to quantify the degree of tacticity by FTIR, with a correlation with solid state nuclear magnetic resonance, NMR. Some important features about crystallinity and structure were concluded by DSC studies. The correlation between DSC and FTIR measurements, along with other techniques, will be carried out in a next phase of this study. There are many possibilities to study PO molecular structures - this study is only the first approach.

REFERENCES

1   Alizadeh, et al., *Polymer*, vol **38**, # 5, 1997, 1207-1214.
2   Balta Calleja et al., *Spectrochimica Acta*, vol **35A**, 1979, 847- 849.
3   Bryant, W.; Voter, R. *Journal of The American Chemical Society*, **75**, 953, 6113-6118.
4   Harvey, M.; Peters, L. *Analytical Chemistry*, **32**,12,1960, 1725.
5   Karger-Kocsis, J. **Polypropylene. Structure, blends and composites. Volume 1: Structure and Morphology**. *Chapman & Hall*, London, 1995, 350.
6   Lomonte, J., *Analytical Chemistry*, **34**,1,1962, 129-131.
7   Mathot, V. **Calorimetry and Thermal Analysis of Polymers**, Hanser, Munich,1994, 353.
8   Mathot. *Journal of Applied Polymer Science*, 1990, **39**, 979.
9   Mitchell, J., **Applied Polymer Analysis and Characterization. Vol. II**. *Hanser*, New York, 1991, 460.
10   Moore et al. **Polypropylene Handbook**, *Hanser*, 1996. Pp 419.

11 Scroder; Muller; Arndt, **Polymer Characterization**. *Hanser*, New York, 1989, 344.
12 Widmann and Riesen. **Thermoanalyse**. *Huethig*, Heildelberg, 1990, 133.

Dynamic Melt Rheometry Used to Study Degradation of Metallocene Polyethylene

Scott H. Wasserman, George N. Foster, Douglas J. Yacka
Univation Technologies, P. 0. Box 670, Bound Brook, New Jersey 08805

INTRODUCTION

THE RELAXATION SPECTRUM INDEX

Dynamic oscillatory shear tests are often used to measure the low shear rheology of polymers. The resulting data can be used to calculate, among other things, the discrete relaxation spectrum. Increasing the breadth of the molecular weight distribution (MWD) or level of long chain branching (LCB) in the polymer system is reflected by an enhanced relaxation spectrum, particularly at high relaxation times. In order to differentiate polymers in terms of their relaxation spectra, we have introduced a quantity known as the Relaxation Spectrum Index (RSI) (see Figure1). The equations used to calculate the RSI are analogous to those used to calculate the polydispersity index (PDI) which describes the breadth of the MWD from size exclusion chromatography (SEC).[1,2] The RSI has proven to be a sensitive indicator of molecular structure, especially where the structure affects molecular entanglements.

MECHANISMS OF POLYMER DEGRADATION

Polyolefins are often exposed to short residence times at high processing temperature (≤320°C) in air or dilute oxygen conditions during fabrication. Either chain breakage, chain branching and/or chain enlargement (or crosslinking) can occur, affecting processing behavior or end-use product performance. Alternately, polyolefin end-use applications can involve extended exposure time in low-temperature (< 150°C), oxidizing environments where chain breakage dominates causing eventual property loss. Often one mechanism dominates as a function of temperature and polymer composition, oxidant type, and concentration.

Illustration of chain breakage and its effect on MWD as measured by SEC for a broad MWD (V catalyzed) PE aged in air at 90°C is given in Figure 2. Anticipated property changes are shown in relation to the progressive MWD change as the polymer undergoes oxidative

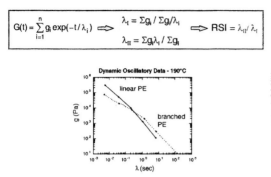

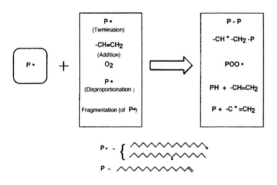

Figure 1. RSI calculation and representative spectra for conventional linear and branched LDPE.

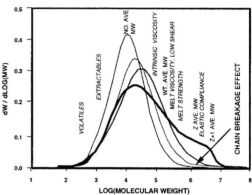

Figure 2. Effect of chain breakage on MWD for a broad MWD (V catalyzed) PE aged in air at 90°C.

degradation. The long polymer chains are being degraded preferentially. To measure such polymer degradation, the MWD change can be measured directly by SEC, or rheological measurements can be used. It is often easier to use measure melt flow changes or rheological shifts with dynamic oscillatory shear rheometry, as will be discussed shortly.

The free radical mechanism for chain branching in PE having vinyl end group unsaturation is summarized in Figure 3. The proposed mechanism leads to three or four-arm star, long chain branching (LCB) configurations that significantly increase low shear viscosity, increasing the low shear melt flow indices such as melt index (MI), $I_{2.1}$. The higher load melt flow indices at somewhat higher shear rates show proportionally less reduction. Therefore, melt flow ratio (MFR), $I_{21}/I_{2.1}$, increases when chain branching dominates over chain breakage.

PE with little or no unsaturations shows predominant chain breakage, especially at high temperatures. As such, it resembles polypropylene (e.g., propylene homopolymer, random propylene copolymers, etc.). This leads, of course, to a MW decrease with breakage of the longer chains being favored statistically. Accordingly, both solution and melt viscosity decrease while MWD narrows and MFR decreases. Again, the presence of even trace amounts of unsaturation (e.g., vinyl-end type) promotes chain enlargement or chain branching (LCB) versus chain breakage. The net effect of chain branching will be a low shear viscosity increase (MI decrease) while intrinsic viscosity will decrease slightly. In addition, PDI corrected for LCB will increase, as will MFR.

RHEOLOGICAL STUDY OF POLYMER STABILITY

For stabilization and product performance control, melt rheological tests can be used to rapidly assess the structural modification of an ethylene polymer. Standard solution characterization techniques such as SEC-MWD may be unable to delineate the effects uncovered by the melt rheology without complicated SEC-coupled viscometry or light scattering systems supported by specialized software. Even then, melt rheology tests are easier to conduct and are probably more sensitive to alterations in the long chain structure. Thus, melt rheology measurements provide a sensitive means for following changes in molecular structure resulting from thermal degradation of the polymer during melt processing. Dynamic rheometry and relaxation spectrum analysis leading to the RSI were recently described.[1]

Low deformation experiments such as dynamic oscillatory shear are considered to be the best for detecting features of molecular structure such as PDI and levels of long-chain branching (LCB). In earlier work, the RSI was used to follow thermo-oxidative degradation during extrusion of unstabilized LLDPE and to quantify its effect.[2] The RSI was shown to be a more sensitive and reliable indicator than the flow activation energy, for example, of the resulting change in molecular structure.

In summary, RSI increase, MI decrease and MFR increase are all related to the development of LCB structures. However, the RSI gives a more sensitive measure of molecular structure changes caused by degradation mechanisms than either the MI or MFR.

EXPERIMENTAL

METHODS

Dynamic oscillatory shear experiments used to calculate the RSI were conducted with a Weissenberg Rheogoniometer manufactured by TA Instruments. Frequency sweep experiments at 190°C and 2% strain in parallel plate mode were run under nitrogen from 0.1 to 100 sec^{-1}. Discrete relaxation spectra were calculated with the commercially available IRIS^{TM} software. RSI is a measure of the breadth of a material's distribution of relaxation times or relaxation spectrum. This dimensionless index has proven to be a sensitive and reliable indicator of long range melt state order (e.g. molecular entanglements) and is a function of polymer MW, MWD, and LCB, as well as temperature.

Melt process stability testing was done using a three pass recycle extrusion method. For the extrusion, the additives in powder form were dry blended with the granular polymer. The

Table 1. Base resin characterization data

Sample	MI	MFR	RSI	RSI.MI$^{0.6}$
Z-N LLDPE	0.96	28	4.5	4.1
mLLDPE A	2.90	18	1.8	3.4
Cr LLDPE	0.75	79	22.0	17.7
HP-LDPE	1.57	61	11.0	14.5
mLLDPE B	1.28	29	8.9	10.3

Sample	Total tertiary H (per 1000C)	Unsaturations (per 1000C)
Z-N LLDPE	19	low
mLLDPE A	17	low
Cr LLDPE	19	high
HP-LDPE	16	moderate
mLLDPE B	11	moderate

method involved the use of a 25 mm diameter single-screw Killion to provide pelleted material using a single strand die, water bath and strand dicer mill. The 1st pass extrusion was done at 210°C; the 2nd and 3rd pass extrusions were done at 260°C. The extruded tape or pellets from each pass were characterized using the test methods noted above. Standard flow indices (i.e., MI and MFR) were measured using ASTM methods.

SAMPLES

The series of five low-density ethylene polymers used in the melt extrusion experiment included Univation Technologies Type I and Type II mLLDPE (mLLDPE A and B, respectively). Ti/Mg or Z-N cat. LLDPE, Cr cat. LLDPE and a HP-LDPE were included to compare their inherent stability and associated stabilization ease to that for the mLLDPEs. The polymers were made in commercial UNIPOL® gas phase PE processes with the exception of mLLDPE B, which is an experimental resin selected for this study because of its higher catalyst residue relative to mLLDPE A and the other LLDPEs. Table 1 summarizes the melt flow, and total tertiary hydrogen and unsaturation for the five base resin polymers. In addition to the RSI, the table lists values for the quantity RSI.MI$^{0.6}$ which effectively adjusts the RSI to a common MI of 1.0.[1] Stabilizers used include primary (hindered phenolic type) and secondary (aryl phosphite ester) antioxidants (PAO-1 and SAO-1) and a catalyst reside deactivator (CRD-1, Zn stearate).[3]

RESULTS AND DISCUSSION

The melt flow and RSI results of the recycle experiments are presented in a series of nine figures (Figures 4-12) using percent change in those properties relative to the five neat or unstabilized polymers. The "P1" and "P3" versions of each polymer refer to samples taken after the first and third passes, respectively. Of the five unstabilized samples presented in Figures 4-6, Z-N LLDPE shows the greatest change in MI, MFR and RSI (normalized to 1.0 MI) even though its unsaturation level is less than either that of the HP-LDPE or Cr cat. LLDPE

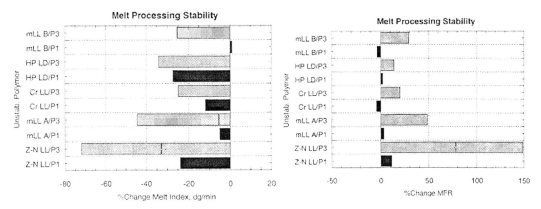

Figure 4. Percentage change in MI for unstabilized samples relative to unprocessed polymer.

Figure 5. Percentage change in MFR for unstabilized samples relative to unprocessed polymer.

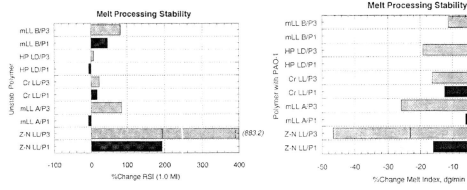

Figure 6. Percentage change in RSI for unstabilized samples relative to unprocessed polymer.

Figure 7. Percentage change in MI relative to unprocessed polymer for samples stabilized with PAO-1.

materials tested. The third pass at 260°C, of course, produces the greatest property shifts. Overall, mLLDPE A and mLLDPE B (high metal residue) show minimal melt flow shift compared to the HP-LDPE and Cr cat. LLDPE. The RSI measurement (see Figure 6) is the most sensitive measure of molecular structure change, indicating that some chain branching is occurring even with the first pass extrusion at 210°C for the Z-N LLDPE (190% increase).

The comparative stabilization response with 0.02 wt.% PAO-1 as presented in Figures 7-9 again shows the Z-N LLDPE to have the more pronounced structure change; however, the change after the third-pass is less than that for the neat polymer. Comparing the percent change of MI, MFR and RSI data after the third pass extrusion, mLLDPE A and B show re-

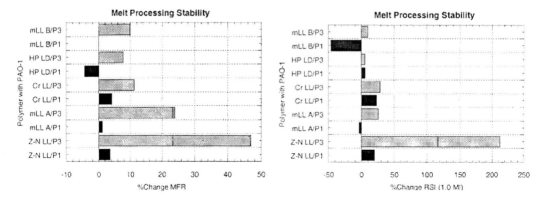

Figure 8. Percentage change in MFR relative to unprocessed polymer for samples stabilized with PAO-1.

Figure 9. Percentage change in RSI relative to unprocessed polymer for samples stabilized with PAO-1.

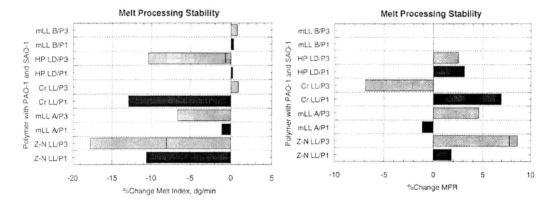

Figure 10. Percentage change in MI relative to unprocessed polymer for samples stabilized with PAO-1 and SAO-1.

Figure 11. Percentage change in MFR relative to unprocessed polymer for samples stabilized with PAO-1 and SAO-1.

sults comparable to the HP-LDPE and Cr cat. LLDPE. Again, the high metal residue mLLDPE B shows comparable stabilization, demonstrating the innocuous nature and state of the metal residues (Zr and Al).

As shown in Figures 10-12, the more potent stabilization system based on 0.02 wt% PAO-1 and 0.08% SAO-1 provides improved protection for most of the five polymers through the 260°C, third-pass extrusion based on the MFR results (< 10% change). However, MI and RSI results indicate that the Z-N LLDPE and perhaps the high-metal-residue mLLDPE B are still showing some LCB after the third-pass. The commercial mLLDPE A, Cr

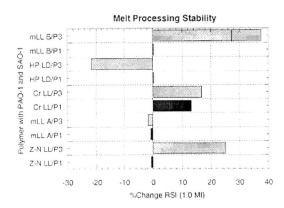

Figure 12. Percentage change in RSI relative to unprocessed polymer for samples stabilized with PAO-1 and SAO-1.

cat. LLDPE and HP LDPE all show evidence of excellent stability and comparable stabilization ease.

CONCLUSIONS

At melt processing temperatures ($>150^{\circ}$C), the presence of even trace amounts of vinyl end group unsaturation promotes chain enlargement or LCB versus chain breakage in either unstabilized or under stabilized polymer compositions. Chain breakage, of course, leads to MW decrease with breakage of the long chains being favored statistically. As the high MW fraction decreases, MI increases, and MWD and MFR decrease. With chain branching predominating over chain breakage, intrinsic viscosity, will decrease. In contrast, low shear viscosity will increase (MI decrease) because of the introduction of LCB. PDI corrected for LCB will increase as will MFR.

Melt rheology measurement provides a sensitive means for following subtle changes in molecular structure resulting from thermal degradation of the polymer as during melt processing. Dynamic oscillatory shear measurements are able to distinguish molecular structure change at low and moderate frequencies for ethylene polymers undergoing thermo-oxidative degradation - i.e., either chain breakage, chain enlargement or chain branching (of a long chain nature). The method is simpler, more straightforward and more sensitive than coupled SEC-viscometry or light scattering methods.

Univation Technologies' mLLDPE having low unsaturation and metal residue which are non-catalytic to the auto-oxidation show excellent intrinsic oxidative stability. Stabilization ease for Univation Technologies' Type I and Type II mLLDPE is equivalent to LLDPE counterparts (similar MI and density) made using high activity, state-of-the-art Cr or Ti/Mg catalyst systems designed for use in the Unipol[®] gas phase PE process.

ACKNOWLEDGMENTS

The fine experimental work done in our rheology and solution characterization laboratories is acknowledged.

NOMENCLATURE

CRD-1	catalyst residue deactivator
g_i	strength of mode in relaxation spectrum
HP-LDPE	high-pressure low density polyethylene
LCB	long chain branch
LLDPE	linear low density polyethylene
MFR	melt flow ratio, $I_{21}/I_{2.1}$
MI	melt index, $I_{2.1}$
mLLDPE	metallocene catalyzed LLDPE
MWD	molecular weight distribution
PAO-1	primary antioxidant
PDI	polydispersity index
RSI	relaxation spectrum index
SAO-1	secondary antioxidant
SEC	size exclusion chromatography
Z-N	Ziegler-Natta
λ_i	relaxation time of mode in relaxation spectrum

REFERENCES

1   S. H. Wasserman, SPE ANTEC Technical Papers, vol. XLIII, Toronto, Canada, p. 1129 (1997).
2   S. H. Wasserman and G. N. Foster, Proc. XIIth Intl. Congress on Rheology, Quebec, Canada, p. 48 (1996).
3   for details, see G. N. Foster, S. H. Wasserman, D. J. Yacka, Proc. 19th Ann. Intl. Conf on Stabilization and Degrad. of Polymers, Lucern, Switzerland, p. 21 (1997).

Melt Rheology and Processability of Conventional and Metallocene Polyethylenes

Choon K. Chai

BP Chemicals s.n.c., Research and Technology Centre, Boîte Postale N° 6, 13117 Lavéra, France

INTRODUCTION

As the dominating deformation in a number of polymer fabrication processes (e.g., in film blowing) is elongational, the Rheotens extensional equipment was developed[1,2] to measure the melt strength and drawability of the polymer melt. It has since proved to be very sensitive to small changes in polymer structure[3,4] and very useful for assessing the bubble stability in film blowing operations.[5] However, until recently, only qualitative interpretation of the Rheotens data could be made as all of these experiments were performed under constant mass flow rate conditions which introduce complex prehistory of the polymer melt in the extrusion die.

Recent developments have shown that Rheotens mastercurves for thermo-rheologically simple polymer melts is dependent only on the extrusion pressure or wall shear stress in the extrusion die and the relative molecular weight distribution (MWD), in variant to changes of melt temperature and the average molecular weight.[6]

In the following, we show that by performing Rheotens experiments at different extrusion pressure, a plot of the melt strength (MS) with extrusion pressure or the corresponding shear rate, allows two derivatives of the melt strength to be defined:

Melt Strength Pressure Derivative $= \partial(MS)/\partial(P)$

Melt Strength Shear Rate Derivative $= \partial(MS)/\partial(\log \dot{\gamma})$

The sensitivity and usefulness of these two MS derivatives, in conjunction with the traditional measured melt strength at a typical extrusion rate, for assessing the processability of a

Table 1. Molecular structure and film blowing performance of PE1 to PE5

Polymer	Type	Melt index 2.16 kg/190°C	M_w/M_n	Motor load A	Melt Temp. °C	Melt pressure bar
PE1	metallocene LLDPE (linear)	1.1	2.1	55	214	275
PE2	metallocene LLD (LC-branched)	1.0	2.2	38	206	176
PE3	Ziegler LLDPE	0.73	8.6	11.1	192	209
PE4	metallocene LLD (LC-branched)	0.77	3.4	12.3	192	199
PE5	autoclave LDPE	1.0	7.0	12.8	189	181

series of conventional and metallocene polyethylenes, as defined by their varying molecular structure (e.g., M_w, MWD and LCB), will be presented.

EXPERIMENTAL

Four LLDPEs (one Ziegler, PE3, and three metallocene, PE1, PE2 & PE4) of similar M_w (~10^5), but their MWD varies from 2.1 to 8.6, are compared (Table 1). The effect of LCB is compared (i) between two narrow MWD metallocenes (PE1 & PE2) and (ii) between a broad MWD metallocene LLDPE (PE4) and a high pressure autoclave LDPE (PE5). The relative influence of broad MWD (of a linear polymer) and LCB (of a narrow MWD branched polymer) on processability (PE3 vs. PE4) is compared; as well as the combined (MWD + LCB) effects of a broad MWD metallocene polymer for achieving high pressure LDPE-like (LDL) processability (PE4 vs. PE5).

Molecular structure parameters, e.g. M_w/M_n the ratio of weight average molecular weight to number molecular weight and g'(LCB) the long chain branching parameter, are characterized by size-exclusion or gel chromatography (SEC or GPC).

The rheological properties of these polymers (Table 1) were characterized at 190°C, using a Rheometrics RDS-2 dynamic rheometer, a Rosand twin-bore (RH7) capillary rheometer and a Göttfert Rheotens extensional rheometer.

The Rheotens extensional experiments measure the melt strength and drawability of the polymer melt (for bubble stability assessment). This is achieved by extruding the polymer at a constant pressure through a die of 1.5 mm diameter and 30 mm in length, with a 90° entry angle. The extrudate is drawn with a pair of gear wheels at an accelerating speed. The drawing force experienced by the extrudate is measured with a transducer and recorded on a chart recorder together with the drawing speed. The maximum force at break are defined as melt strength (MS) of the polymer melt, for a given extrusion temperature (190°C) and pressure.

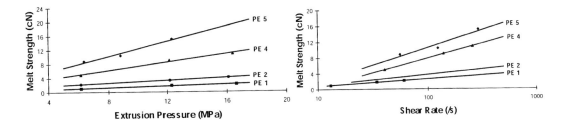

Figure 1. Processability. Melt strength at 190°C.   Figure 2. Processability. Melt strength at 190°C.

Film blowing performance of PE1 and PE2 were evaluated on a 50 mm Reifenhauser extruder with a die diameter of 150 mm, die gap of 2.3 mm, BUR of 2:1 and a constant output of 50 kg/h, for a 38 μm thick film. PE3 to PE5 were blown on a 45 mm Collin extruder with a die diameter of 100 mm, die gap of 0.8 mm, BUR of 3:1 and a constant output of 13 kg/h.

RESULTS AND DISCUSSION

The MS-Pressure and MS-(Shear Rate) plots of Figures 1 and 2 provide very useful information regarding the processability of the polymer: e.g. for the film blowing process - good balance between extrusion melt pressure and bubble stability is required. This novel approach represents an advancement over the traditional method of measuring just a single melt strength curve at a fixed shear rate. These figures clearly demonstrate and distinguish the various types of polyethylenes in terms of their processability: (i) poor processability typical of a narrow MWD metallocene LLDPE (e.g. PE1), (ii) easy processing narrow MWD long chain branched, metallocene LLDPE (e.g. PE2), (iii) excellent processing typical of a high pressure LDPE (e.g. PE5) and (iv) low pressure LDPE-like (LDL) processability with broad MWD long chain branched metallocene LLDPE (e.g. PE4). These differences can be quantified, apart from the traditional melt strength measured at a typical extrusion rate, e.g. of 500 s^{-1}, MS(500 s^{-1}) as determined from Figure 2, with the slopes of these plots: $\partial(MS)/\partial P$ in Figure 1 and $\partial(MS)/\partial(\log \dot{\gamma})$, in Figure 2, which yield some of the very important processing and molecular structure information. These two melt strength derivatives, $\partial(MS)/\partial P$ and $\partial(MS)/\partial(\log \dot{\gamma})$, together with MS(500 s^{-1}) and its associated extrusion pressure, P(500 s^{-1}) form the essential information needed to characterize the processability of a polymer, if the flow is stable or the instability duration is outside the processing window. The melt strength derivatives are strongly dependent on MWD and LCB while MS(500 s^{-1}) and P(500s^{-1}) vary

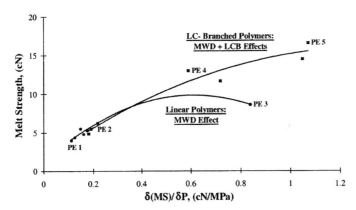

Figure 3. Processability. Bubble stability at 190°C.

with molecular weight, MWD and LCB. The dependence of the latter include its concentration, type and distribution within the MWD.

To illustrate this new concept, Figure 3 shows the plot of MS(500 s^{-1}) against ∂(MS)/∂P for PE1 to PE5, and other polymers not shown in Table 1, with melt index ~ 1.

In general, it is desirable for a polymer to be at the right top corner (e.g. PE5 - high pressure LDPEs exhibit high

∂(MS)/∂P values and hence their excellent bubble stability and extrudability balance characteristics) rather than at the lower left hand corner (e.g. PE1), though too high a MS(500 s^{-1}) value may incur drawability (i.e. drawdown potential) problem or limitation in some polymer processing applications. However, high value of ∂(MS)/∂P is very desirable as it signifies the degree of improvement in bubble stability per unit (MPa) increase or penalty in extrusion melt pressure.

To illustrate this point further, the ∂(MS)/∂P for PE5 is ~1 cN/MPa as compared to ~ 0.1 cN/MPa for PE1, indicating that the extrusion melt pressure of PE1 will increase ~ 10 times faster than PE5 for an improvement of 1 cN in melt strength (i.e. bubble stability). That is the processability of PE1 will be melt pressure (and melt fracture) limited well ahead of PE5. In particular, PE5 exhibits higher melt strength at a typical processing rate, i.e. MS(500 s^{-1}) ~ 16 cN as compared to ~ 4 cN for PE1. Their corresponding extrusion pressures, P(500 s^{-1}) are 14 and 30 MPa respectively. Thus, it will be practically impossible for PE1 to achieve the same melt strength level as PE5, as it requires a corresponding melt pressure of ~150 MPa! In fact PE1 exhibits shark-skin and melt fracture at melt pressure below ~ 30 MPa. This new method clearly demonstrates the expected poor processability for the standard, linear narrow MWD metallocene LLDPEs with no LCB.

MOLECULAR WEIGHT DISTRIBUTION (MWD) EFFECT

One approach to improve processability is by broadening the MWD. As respectively shown in Figures 3 and 4, both melt strength and its pressure derivative ∂(MS)/∂P for the linear polymers increase as their MWD broadens: e.g. PE1 to PE3. However, these improvements do not continue indefinitely: both MS(500s^{-1}) and ∂(MS)/∂P reach a maximum and then begin to de-

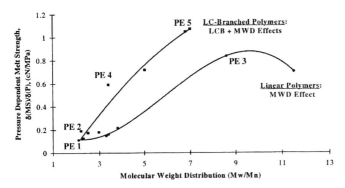

Figure 4. Molecular weight distribution effect.

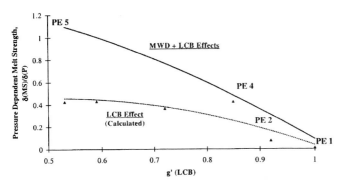

Figure 5. Long chain branching effect.

crease with MWD. This may be due to the molecular chain "dilution" effect, which appears to become significant at MWD > ~ 7. In addition, the mechanical and optical properties are known to deteriorate as MWD is broadened.

LONG CHAIN BRANCHING (LCB) EFFECT

An alternative approach, which has been shown to be more effective in improving processability of LLDPE (both extrudability and bubble stability) while maintaining the mechanical and optical properties of the polymer, is the inclusion of LCB. This is clearly demonstrated in Table 1 where the film blowing performance of the long-chain branched narrow MWD metallocene polymer (PE2) is much superior than its linear counterpart (PE1) in terms of lower motor load, melt temperature and pressure.

Figures 3 and 4 showed that both MS(500 s^{-1}) and ∂(MS)/∂P values for the easy processing PE2 are higher than those of PE1. In addition, the lower curve in Figure 5 shows the effect of LCB on ∂(MS)/∂P, which is calculated by subtracting the MWD influence on the long chain-branched samples using the data in the lower curve of Figure 4 for the linear polymers where their ∂(MS)/∂P are due mainly from the MWD effect. It shows that ∂(MS)/∂P increases with LCB as measured by GPC (g'$_{LCB}$ value).

COMBINED MWD AND LCB EFFECTS: LOW DENSITY POLYETHYLENE-LIKE (LDL) PROCESSABILITY

To achieve the conventional high pressure low-density polyethylene-like (LDL) processability (e.g. that of PE5), it appears that broadening MWD alone is not sufficient (e.g., PE3). This is clearly demonstrated in Table 1, Figures 3 and 4 where with a MWD ~ 9, the melt pressure and bubble stability-extrudability balance of PE3 is not nearing that of the LDPE (PE5).

Figures 3, 4 & 5 show that the excellent processability of the LDPE (PE5) is achieved with a broad MWD and the presence of LCB. It appears that LCB roughly doubles the $\partial(\mathrm{MS})/\partial P$ values for a given MWD (upper curves of Figures 4 and 5) and extends the melt strength values for polymers of similar melt index (upper curve of Figure 3).

The improvement in processability of sample PE4 over PE2 is achieved with the combined MWD and LCB effects. As shown in Figures 3 to 4, the combined effects have just approached about mid-way the processability of PE5, thus the film blowing performance of PE4, as shown in Table 1, is better than PE3 (improvement by MWD broadening effect only) but, as expected, is not similar to that of PE5. However, it is clear that the LDPE-like processability is achievable, by broadening the MWD of the long chain branched metallocene LLDPEs and by modifying the concentration, type and distribution of LCB, as indicated by the directions of the upper curves of Figures 3, 4 & 5.

CONCLUSIONS

An improved Rheotens melt strength test, based on measurements performed at different extrusion pressures, has been developed. It has been used to investigate the processability, in particular, the bubble stability and extrudability balance, of a series of conventional and metallocene polyethylenes. Two new parameters: $\partial(\mathrm{MS})/\partial P$, the melt strength pressure derivative and $\partial(\mathrm{MS})/\partial(\log \dot{\gamma})$, the melt strength shear rate derivative, have been defined.

These two melt strength derivatives, together with melt strength measured at a typical extrusion rate, e.g. of 500 s^{-1}, MS(500 s^{-1}) and its associated extrusion pressure, P(500 s^{-1}) form the essential information needed to characterize the processability of a polymer, if the flow is stable or the instability duration is outside the processing window. The sensitivity of these parameters to features of molecular structures (e.g. M_w, MWD and LCB) make them best suited for characterizing the processability of entire polyethylene families.

For example, the excellent processability of LDPE is characterized by its high $\partial(\mathrm{MS})/\partial P$, $\partial(\mathrm{MS})/\partial(\log \dot{\gamma})$ and MS(500 s^{-1}) values, and a low value in its associated extrusion pressure P(500 s^{-1}). Conversely, the poor processability of the standard narrow MWD metallocene LLDPE exhibits low values in melt strength and its two derivatives but high extrusion pressure P(500 s^{-1}).

The LDPE-like processability is achievable, by broadening the MWD of the long chain branched metallocene LLDPEs and by modifying the concentration, type and distribution of LCB.

ACKNOWLEDGMENTS

The author thanks Mr René Giraud for the rheological measurements, colleagues for useful discussion and BP Chemicals for permission to publish this work.

REFERENCES

1   J. Meissner, Deutsche Offenlegungsschrift DOS 1904079(1969).
2   J. Meissner, *Rheol. Acta*, **10**, 230-242 (1971).
3   J. Meissner, Basic Parameters, Melt Rheology, Processing and End-use Properties of Three Similar Low Density Polyethylene Samples, *Pure Appl. Chem.*, **42**, 553-612 (1975).
4   F. P. LaMantia and D. Acierno, Influence of Molecular Structure on the Melt Strength and Extensibility of Polyethylenes, *Polym. Eng. Sci.*, **25**, 279-283 (1985).
5   A. Ghijsels, J. J. S. M. Ente and J. Raadsen, Melt Strength Behavior of PE and Its Relation to Bubble Stability in Film Blowing, *IPP*, **5**, 284-286 (1990).
6   M. H. Wagner, V. Schize and A. Goettfert, Rheotens-Mastercurves and Drawability of Polymer Melts, *Polym. Eng. Sci.*, **36**, 925-935 (1996).

Rheology of Metallocene-Catalyzed Polyethylenes.
The Effects of Branching

Sujan E. Bin Wadud and Donald G. Baird
Department of Chemical Engineering, Virginia Polytechnic Institute and State University,
Blacksburg, VA 24061-0211

INTRODUCTION

It is known that LDPE, followed by HDPE and then LLDPE, is the most stable in the film blowing process.[1] The stability is related to the extensional viscosity exhibited in general by the three different types of polyethylenes (PE's). In particular, LDPE which is highly branched exhibits a significant degree of extensional strain-hardening, whereas LLDPE exhibits none and HDPE can exhibit some as a result of a broad molecular weight distribution (MWD).[2] LDPE also exhibits an onset of shear-thinning at very low shear rates, usually less than 0.1 s^{-1} whereas LLDPE does not shear-thin readily. With the advent of metallocene-catalyzed polyethylenes (m-PE's) some of the problems associated with LLDPE such as processing stability and the lack of shear thinning, readily arise as a result of the narrow MWD possible with these systems.[3] In an effort to improve on the processing performance of m-PE's, researchers at Dow[4] have indicated that branching can be incorporated into m-PE's by forming copolymers containing an alpha-olefin such as octene. However, it is not clear what the effect of branching is on the flow behavior of these systems. There is some indication that the addition of a few long chain branches can promote shear-thinning behavior. Furthermore, it is extremely difficult to identify branching analytically without assuming that it is present. The purpose of this paper is to present studies concerned with the extensional and shear viscosity behavior, supercooling effect, and other stress-strain relationships of three PE's which have apparently different degrees of long chain branching but similar melt flow indexes.

Table 1. Molecular weight, molecular weight distribution, zero shear viscosities and branching of four metallocene-catalyzed polyethylenes

Sample	MFI*	Molecular weight x10^{-3}				η_o** Pa.s	η_o** Pa.s	LCB***
		M_n	M_w	M_z	M_w/M_n			
Dow Affinity PL 1880	1	33.4	77.1	143	2.31	10500	-	0.4240
Dow Affinity PL 1840	1	36.3	87.9	160	2.42	47000	15700	0.5695
Dow Affinity PL 1850	(3)?	55.1	116.4	183	2.11	46000	17200	0.1773
Exxon Exact 3022	1	45.3	95.8	165	2.11	16000	6500	no LCB

*reported; **at 190°C; ***LCB=long-chain branches/10,000 carbon;
Note: MW and branching data provided by Wilem deGroot, Dow Chemical Co.

EXPERIMENTAL

The metallocene-catalyzed polyethylenes used were Dow Affinity PL 1880, Affinity PL 1840 and Affinity PL 1850, which are copolymers of ethylene and octene-1, and Exxon Exact 3022. The Dow polymers are branched copolymers with controlled degrees of branching (Table 1) and narrow MWD (Figure 2). The Exxon LLDPE does not have LCB.

SAMPLE PREPARATION AND TESTING

Samples for dynamic shear rheological testing were prepared on the heated plates of the rheometer by placing polymer pellets on the lower plate and then bringing the higher plate to press the molten polymer into a flat disk about 1.4 mm in thickness. The Rheometrics Mechanical Spectrometer (RMS 800) with 25 mm diameter parallel plates was utilized for dynamic testing, and 25 mm plates with 0. 1 rad. cone for all steady shear experiments. The test sample was maintained in a nitrogen atmosphere during all tests. A probe that was in contact with the lower plate monitored the sample temperature. All tests were conducted at 150±0.1°C.

Complex viscosity measurements were made in small amplitude oscillatory testing mode over angular frequency range of 0.04 to 100 rad/s. The steady shear experiments performed at shear rates 0.001 to 0.1 s^{-1}.

The hysteresis curves (Figure 7) were generated by subjecting each sample to different steady shear rates at 150°C and recording the equilibrium stress that was built up over time. An upward sweep was performed that was immediately followed by a downward sweep.

DSC scans were performed with a Seiko Instruments DSC 220C with liquid nitrogen autocooler unit. All results are normalized to 1 mg sample weight.

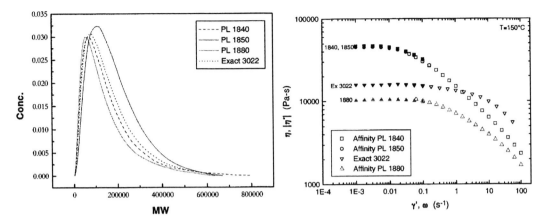

Figure 1. Molecular weight distribution of PL 1840, PL 1850, PL 1880, and Exact 3022.

Figure 2. Flow curves of PL 1840, PL 1850, PL 1880 and Exact 3022 at 150°C.

Cylindrical samples of the polymers were molded using a molding fixture. The mold was heated to 150°C and polymer pellets were dropped in. Vacuum was pulled through the system in order to avoid polymer degradation in air. A plunger was used to pack the molten polymer in a hydraulic press at about 2270 kg (5000 lbs) and then cooled at room temperature. The cylinder diameter ranged from 5.72 mm to 5.79 mm. These samples were glued to clips using two-part epoxy adhesive (EPOXI-PATCH, Dexter Adhesives and Structural Materials). Constant extension rate uniaxial elongational tests were performed using Rheometrics Extensional Rheometer (RER 9000). The test sample remained immersed in a neutrally buoyant oil bath whose temperature was maintained at 150 ± 0.1°C. This temperature was selected because there was uniform deformation along the length of the sample with minimum necking.

RESULTS AND DISCUSSION

In general, the metallocene-catalyzed PE's are known to have narrow MWD's (Figure 1) which is reflected in their viscosity behavior (Figure 2). The filled points in Figure 2 were obtained from steady-shear experiments using cone-and-plate geometry, and the open points from dynamic oscillatory tests using parallel plates. In particular, sample Ex3022 exhibits a long Newtonian plateau, i.e. the viscosity stays constant up to high values of ω (which is equivalent to high shear rates by Cox-Merz rule[5]) and only barely begins to exhibit shear-thinning behavior. On the other hand, the two branched samples PL1840 and PL1850

Table 2. Melting point and activation energies of metallocene-catalyzed PEs

Material	Melting point, °C	ΔE_a, K	ΔE_a, kJ/mol	ΔE_a, kcal/mol
Exxon Exact 3022	112	3240±310	27.0±2.5	6.4±0.6
Dow Affinity PL 1840	108	4200±244	35.0±2.0	8.3±0.4
Dow Affinity PL 1850	103	4330±252	36.0±2.0	8.6±0.5

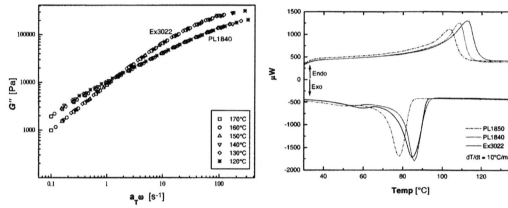

Figure 3. Mastercurves of G′ for PL 1840 and Exact 3022 at T_{ref}=170°C.

Figure 4. DSC heating and cooling curves of PL 1840, PL 1850 and Exact 3022.

exhibit higher zero shear viscosities (Table 1) and an earlier onset of shear-thinning. It is interesting that even though Ex3022 has higher MW than PL1840, its zero-shear viscosity is much lower (by 80%, Table 1). Again, branching and MW seem to be having opposite effects on zero-shear viscosity for PL1840 and PL1850, because despite the fact that the former polymer has a much lower MW, its zero-shear viscosity is almost identical to that of PL1850. This higher viscosity may be attributed to either its higher LCB content or higher MWD. At higher frequencies, the viscosity curves of PL1840 and PL1850 do seem to separate out which suggests that the two polymers have similar relaxation behavior at long times only.

The activation energy of PL1840, PL1850 and Ex3022 are shown in Table 2. These were obtained by measuring G″ at different temperatures (Figure 3), and then shifting them to 170°C using time-temperature superposition principle.[6,7] As can be expected, the activation energy of the linear polymer, namely Ex3022, is lower than its branched counter-part with similar MW, PL1840 (Table 2). There are no discernable differences in activation energy of PL1840 (35±2 kJ/mol) and PL1850 (362 kJ/mol).

Melting points of PL1840, 1850 and Exact 3022 are shown in Table 2. DSC diagrams in Figure 4 show that in the cooling cycle, there is evidence of a "soft" crystallization peak at

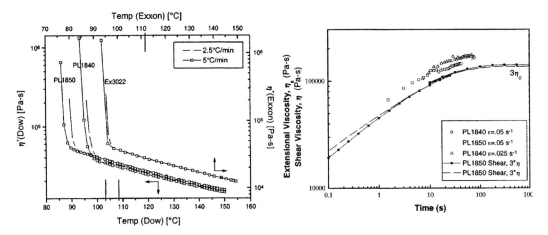

Figure 5. Cooling curves of PL 1840, PL 1850, and Exact 3022 at 2.5 and 5°C/min lines on temperature axis indicate melting points.

Figure 6. Transient extensional viscosity and dynamic viscosity at 150°C.

lower temperatures. This is more pronounced in Ex3022 around 60°C than the Dow materials. Hill *et al.*[8] concluded that such a broad peak is due to the phase-separation of low MW material and high MW fractions. This "soft" peak is more pronounced for conventional LLDPE's which have heterogeneous branching distributions, as observed by Gabriel *et al.*[9]

Temperature sweeps (Figure 5) at 2.5 and 5°C/min show that supercooling is enhanced in PL 1840 and PL 1850 and not so in Ex3022. When the sweep is conducted at a faster rate, the branched polymers are able to remain in the molten state at lower temperatures, but not so for linear Ex3022. This suggests that the presence of long-chain branches may be hindering the nucleation process, i.e. the initiation step toward crystallization.

Under extension, as shown in Figure 6, PL1840 exhibited some strain-hardening whereas PL1850 behaved more like the linear systems. In fact, even though the shear viscosity is highest for sample PL1850, the extensional viscosity of sample PL1840 rises above that of sample PL1850, more so at lower extension rates. These results support the fact that sample PL1840 has more number of branches/10,000 carbon than sample PL1850 (Table 1).

Another interesting finding is that when the branched polymer PL1840 was subjected increasing shear rates, the path that the equilibrium stress takes is different from that when subjected to decreasing shear rates (Figures 7 & 8). The filled squares are equilibrium stresses for upward rate sweep, and the open circles, the downward sweep. Hence, there seems to be a stress-strain hysteresis effect for PL1840. But this is not the observed in the case of Ex3022 (Figure 8), which has the similar MW as PL1840. The linear material retraces its way back to the lower shear rate stress value almost exactly the way as the stress grew. Hence, the with the

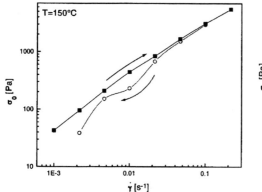

Figure 7. Stress-strain hysteresis curve of PL 1840.

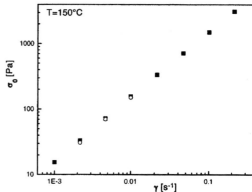

Figure 8. Stress-strain hysteresis curve for Exact 3022.

presence of LCBs, the polymer is unable to retain its initial unstressed state completely. This effect was reproducible to within 2% relative error.

CONCLUSIONS

The strain-hardening present in at least one of the samples supports the presence of long chain branches. The shear flow properties of the two samples which are apparently branched cannot be used by themselves as indicators of branching, as seen for PL1840 and PL1850, but must be combined with extensional flow measurements. DSC shows the presence of a short-chain, low molecular weight phase within the melt, especially for linear metallocene-catalyzed PE. Stress strain hysteresis suggests that branching plays an important role in loss of original morphology and its inability to recover stress. The presence of branching also reduces the crystallization rates.

Additional studies on the flow of these materials in the contraction region of a die along with their melt fracture behavior should be useful in distinguishing whether small changes in the molecular architecture will noticeably affect their processing performance. Furthermore, there is a need to analyze the relaxation spectrum of these materials in order to ascertain how the presence of high MW tails and branching manifest themselves in the relaxation strengths.

ACKNOWLEDGMENTS

We would like to thank Dow Chemical Company, Midland, Michigan, for supplying the polyethylene resins. Exxon Plastics for supplying the Exact 3022. Dr. Willem DeGroot for providing the MW, MWD and branching information. Dr. Garth Wilkes, Virginia Tech, for the use of his DSC. Varun Ratta for running the DSC scans.

REFERENCES

1   W. Minoshima and J. L. White, *J. Non-Newt. Fluid Mech.*, **19**, 275-302 (1986).
2   H. Muenstedt and H. Laun, *Rheol. Acta*, **20**, 211-221 (1981).
3   J. F. Vega, A. Munoz-Escalona, A. Santamaria, M. E. Munoz, and P. Lafuente, *Macromolecules*, **29**, 960-965(1996).
4   Y. S. Kim, C. I. Chung, S. Y. Lai, and K. S. Hyun, SPE Technical Papers, 41, 1122-1128(1995).
5   E. H. Cox and W. P. Merz, *Journal of Polymer Science*, **28**, 619-621 (1958).
6   J. D. Ferry, **Viscoelastic Properties of Polymers**, 2ed, 1970.
7   D. G. Baird and A. Collias, **Polymer Processing**.
8   M. J. Hill, P. J. Barham, J. van Ruiten, *Polymer*, **34**(14), 2975-2980 (1993).
9   C. Gabriel, J. Kaschta and H. Muestedt, copy of manuscript submitted to Rheologica Acta.

Extensional and Shear Rheology of Metallocene-Catalyzed Polyethylenes

Sujan E. Bin Wadud and Donald G. Baird
Department of Chemical Engineering, Virginia Polytechnic Institute and State University,
Blacksburg, VA 24061-0211

INTRODUCTION

It is known that LDPE, followed by HDPE and then LLDPE, is the most stable in the film blowing process.[1] The stability is related to the extensional viscosity exhibited in general by the three different types of polyethylenes (PE's). In particular, LDPE which is highly branched exhibits a significant degree of extensional strain-hardening, whereas LLDPE exhibits none and HDPE can exhibit some as a result of a broad molecular weight distribution (MWD).[2] LDPE also exhibits an onset of shear-thinning at very low shear rates, usually less than 0.1 s^{-1} whereas LLDPE does not shear-thin readily. With the advent of metallocene-catalyzed polyethylenes (m-PE's) some of the problems associated with LLDPE such as processing stability and the lack of shear thinning, readily arise as a result of the narrow MWD possible with these systems.[3] In an effort to improve on the processing performance of m-PE's, researchers at Dow[4] have indicated that branching can be incorporated into m-PE's by forming copolymers containing an alpha-olefin such as octene. However, it is not clear what the effect of branching is on the flow behavior of these systems. There is some indication that the addition of a few long chain branches can promote shear-thinning behavior. Furthermore, it is extremely difficult to identify branching analytically without assuming that it is present. The purpose of this paper is to present studies concerned with the extensional and shear viscosity behavior, supercooling effect, and other stress-strain relationships of three PE's which have apparently different degrees of long chain branching but similar melt flow indexes.

EXPERIMENTAL

MATERIALS

The metallocene-catalyzed polyethylenes used were Dow Affinity PL 1880 (65), Affinity PL 1840 (66) and Affinity PL 1850 (67), which are copolymers of ethylene and octene-1, and Exxon Exact 3022 (68). The Dow polymers are branched copolymers with controlled degrees of branching and narrow MWD. The Exxon LLDPE is unbranched.

SAMPLE PREPARATION AND TESTING

Samples for dynamic shear rheological testing were prepared on the heated plates of the rheometer by placing polymer pellets on the lower plate and then bringing the higher plate to press the molten polymer into a flat disk about 1 mm in thickness. The Rheometrics Mechanical Spectrometer (RMS 800) with 25 mm diameter parallel plates was utilized. The test sample was maintained in a nitrogen atmosphere during all tests. A probe that was in contact with the lower plate monitored the sample temperature. Complex viscosity measurements were made in small amplitude oscillatory testing mode over angular frequency range of 0.04 to 100 rad/s. All tests were conducted at 150±0.1°C.

Cylindrical samples of the polymers were molded using a molding fixture. The mold was heated to 150°C and polymer pellets were dropped in. Vacuum was pulled through the system in order to avoid polymer degradation in air. A plunger was used to pack the molten polymer in a hydraulic press at about 2270 kg (5000 lbs) and then cooled at room temperature. The cylinder diameter ranged from 5.72 mm to 5.79 mm. These samples were glued to clips using two-part epoxy adhesive (EPOXI-PATCH, Dexter Adhesives and Structural Materials). Constant extension rate uniaxial elongational tests were performed using Rheometrics Extensional Rheometer (RER 9000). The test sample remained immersed in a neutrally buoyant oil bath whose temperature was maintained at 150±0.1°C. This temperature was selected because there was uniform deformation along the length of the sample with minimum necking.

RESULTS AND DISCUSSION

As a point of reference the extensional stress versus strain of a conventional LDPE (c-LDPE) is compared to that of conventional LLDPE(c-LLDPE) in Figure 1. The temperatures were selected so that the shear viscosity of c-LLDPE was similar to that of c-LDPE. There is a distinct difference in the growth of extensional stress for the two PE's. In particular, for c-LLDPE the stress rises rapidly to steady state and tends to level off. The rapid rise to steady state is due to the short relaxation times which are related to a somewhat narrow MWD for the c-LLDPE. On the other hand, the extensional stress of c-LDPE rises more slowly with

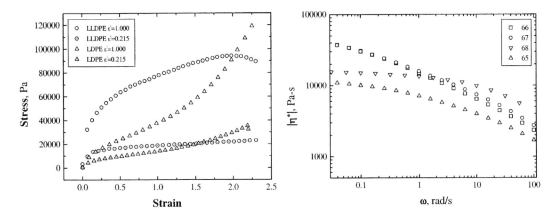

Figure 1. Stress vs. strain curve for LDPE at 170°C and LLDPE at 150°C at constant strain rates 0.215 and 1.00.

Figure 2. Complex viscosity vs. frequency for samples 65, 66, 67 and 68 at 150°C over frequency range 0.04 to 100 rads/s.

strain(and hence time) but continues to rise above that for c-LLDPE. This increase of extensional stress with increasing strain is referred to as strain-hardening. The transient response of the highly branched PE is related to a broad MWD which provides longer relaxation times and the attempt of the fluid to rise to a high steady state stress value, which is related to the presence of long chain branches. Hence, even though the shear viscosity of these two PE's is similar at these temperatures, the extensional viscosity of the branched PE is higher.

Before looking at the extensional viscosity of the metallocene-catalyzed PE's(m-LLDPE), we investigate their dynamic shear behavior to provide a common point of reference. The metallocene-catalyzed PE's are known to have narrow MWD's which is reflected in their viscosity behavior. In particular, as shown in Figure 2 for sample 68, the viscosity stays constant up to high values of ω (which is equivalent to high shear rates) and only barely begins to exhibit shear-thinning behavior. On the other hand, the two branched samples (66 and 67) exhibit higher zero shear viscosities and an earlier onset of shear-thinning. Although it is thought that this behavior is related somehow to the presence of long chain branching, these changes could also be due to changes in MW and MWD. However, as observed in Figure 2, $|\eta^*|$ at 150°C for samples 66 and 67 are nearly identical which suggests MW and MWD are very similar for the two PE's. What is more intriguing is the fact that at 190°C (Figure 3) the shape of the flow curves for the two branched samples resemble each other. In fact the data can be superimposed on each other by shifting mostly vertically

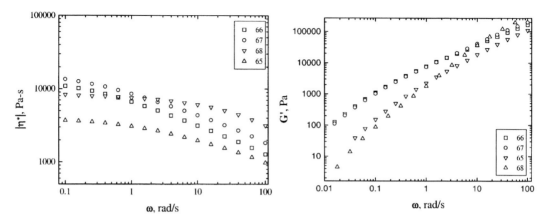

Figure 3. Complex viscosity vs. frequency for samples 65, 66, 67 and 68 at 190°C over frequency range 0.1 to 100 rad/s.

Figure 4. Storage modulus vs. frequency for samples 65, 66, 67 and 68 at 150°C over frequency range 0.04 to 100 rad/s.

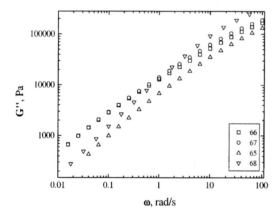

Figure 5. Loss modulus vs. frequency for samples 65, 66, 67 and 68 at 150°C over frequency range 0.0.4 to 100 rad/s.

along the viscosity axis and only slightly along the shear rate axis. This suggests that the activation energies are different for the tow samples.

On the other hand, further evaluation of the dynamic data by evaluating the elastic (G') and viscous(G") components of the response separately (Figures 4 and 5, respectively) reveals some interesting results. At angular frequencies less than about 10 rad/s, G' of samples 66 and 67 is nearly identical while for sample 68 G' is much lower. At frequencies greater than 10 rad/s the values of G' are certainly highest for sample 68. In fact there is more difference between the samples 66 and 67 at high frequencies than at low frequencies, which means that there are more differences in the short relaxation times than the long relaxation times. However, for samples 66 and 67 which are supposedly branched, the effect on the long relaxation times is most noticeable relative to the linear sample 68. In contrast to the G' behavior, G" of samples 66, 67, and 68 is somewhat similar at low frequencies, but deviate from each other more at high frequencies. In other words if the differences were due to the magnitude of MW,

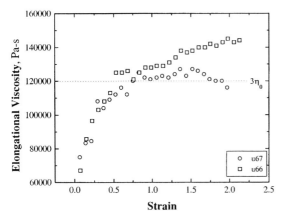

Figure 6. Elongational viscosity vs. strain curve for samples 66 and 67 at 150°C at constant strain rate 0.05.

then we might expect the behavior of G" to resemble more closely that of G'. In any event the dynamic response is somewhat contradictory to having just changes in MW.

The most revealing results seem to be in examining the extensional viscosity of the samples. The extensional viscosity for sample 68, basically a linear PE of narrow MWD, which is not shown, is basically similar to that shown for LLDPE. In particular, the extensional viscosity rises rapidly to steady state and because of the lack of strain-hardening the samples tended to neck and become unstable on extension. However, as shown in Figure. 5, sample 66 exhibited some strain-hardening whereas sample 67 behaved more like the linear systems. In fact, even though the shear viscosity is highest for sample 67, the extensional viscosity of sample 66 rises above that of sample 67. These results suggest that sample 66 has more branches than does sample 67. Although the extensional behavior of sample 67 more resembles that of a linear polymer or at least one with a few short chain branches, there are still subtle signs of branching present.

CONCLUSIONS

The strain-hardening present in at least one of the samples is very suggestive of the presence of long chain branches. The shear flow properties of the two samples which are apparently branched can not be used by themselves as indicators of branching but must be combined with extensional flow measurements. Additional studies on the flow of these materials in the contraction region of a die along with their melt fracture behavior should be useful in distinguishing whether small changes in the molecular architecture will noticeably affect their processing performance.

ACKNOWLEDGMENTS

We would like to thank Dow Chemical Company, Midland, Michigan, for supplying the polyethylene resins.

REFERENCES

1   W. Minoshima and J. L. White, *J. Non-Newt. Fluid Mech.*, **19**, 275-302(1986).

2   H. Muenstedt and H. Laun, *Rheol. Acta*, **20**, 211-221(1981).
3   J. F. Vega, A. Munoz-Escalona, A. Santamaria, M. E. Munoz, and P. Lafuente, *Macromolecules*, **29**, 960-965(1996).
4   Y. S. Kim, C. I. Chung, S. Y. Lai, and K. S. Hyun, SPE Technical Papers, 41, 1122-1128(1995).

A Criterion for the Onset of Gross Polyolefins: Molecular Structure Effect

Seungoh Kim and John. M. Dealy
McGill University, Montreal, Canada

INTRODUCTION

In polymer melt extrusion, gross melt fracture, generally characterized by a chaotic distortion of the extrudate, is often observed when the extrusion rate is increased beyond a certain value. Since gross melt fracture limits the production rate, much attention has been devoted to understanding its origin. However, there is still no general understanding of the origin of this phenomenon.[3] Most of the previously proposed mechanisms and criteria for the onset of gross melt fracture (OGMF) are based on quantities calculated assuming shear flow ; e.g., a critical value of apparent wall shear rate or shear stress at the wall in capillary flow. These criteria, while useful in comparing similar materials under the same flow conditions, sometimes give inconsistent results when comparing data from different temperatures and die geometries.[1] They are also affected by slip at a certain stress level.

Moreover, correlating the onset of gross melt fracture with shear flow quantities has not provided a satisfactory correlation of experimental observations. Direct visualization studies and streaming birefringence patterns at the entrance of a capillary indicate that gross melt fracture is initiated at the capillary inlet, where the fluid deformation is primarily an extensional flow.[5] This implies that extensional flow plays an important role in gross melt fracture.

From a practical point of view, it is more important to establish a reliable criterion for OGMF than to formulate a complete theory of its origin. The objective of this paper is not concerned directly with the origin of gross melt fracture, but the conditions under which gross melt fracture will be expected to occur.

A criterion for the onset of gross melt fracture based on extensional properties is proposed. Extensional properties are estimated from the entrance pressure drop using the analyses of Cogswell[1] and Binding.[2] The effects of molecular weight (MW), molecular weight distribution (MWD) and degree of long chain branching (LCB) on the critical extensional stress are presented.

THEORY AND BACKGROUND

Several attempts have been made to correlate melt fracture with tensile failure. Denn *et al.*[6] presented a theoretical analysis and experimental data indicating that there is a maximum stretch ratio to which a polymeric material can be subjected. Cogswell[7] suggested the existence of a maximum tensile stress for polymer melts by using converging die technique. Recently Ghijsels *et al.*[8] also presented tensile stress failure data for many polymer melts using a Rheotens apparatus.

There have also been other attempts to correlate the extensional flow component with gross melt fracture by using entrance pressure drop data. Everage *et al.*[9] proposed a critical recoverable extension rate as a criterion. Shaw[10] suggested a critical tensile stress for OGMF by using a lubricated conical die, while Hürlimann *et al.*[11] proposed a critical entrance pressure drop. These approaches are interesting, but oversimplified. For example, Hürlimann[11] ignored the shear component and assumed that the total entrance pressure drop in an orifice die is due to tensile stress. Everage *et al.*[9] derived a criterion assuming that the contraction angle is equal to the entrance angle of the die. But this assumption and the resulting criterion have not been verified. Recently, Gibson[12] and Kwag *et al.*[13] showed that the ratio of shear to extensional contributions to the pressure drop can change, depending on the geometry and shear rate. For example, in a converging die having a conical. entrance angle, the contribution of the shear component to the total pressure drop can be neglected if the entrance angle is greater than 60 degrees.

CONVERGING FLOW ANALYSIS

A converging die provides a convenient method for studying extensional flow. Converging die flow is easy to generate by use of a capillary rheometer and is dominated by extensional deformation. A number of approximate analyses have been proposed to relate the entrance pressure drop of converging die flow to an extensional viscosity. Cogswell[1] made an analysis by considering both shear and extensional flow components. Gibson[12] modified Cogswell's analysis, but assumed sink flow kinematics. Binding[2] followed Cogswell's idea but presented a more rigorous analysis by using variational principles to minimize the energy dissipation in the contraction instead of using Cogswell's pressure drop minimization hypothesis.

Recently' Padmanabhan *et al.*[14] and Mackay *et al.*[15] compared Cogswell's and Binding's methods and showed that both models are valid at high extension rates. Since gross melt fracture occurs in this zone, both methods seem to be acceptable for inferring the extensional properties of the flow.

Table 1. Molecular parameters of CGC, mPE and LLDPE

Resin	Density^a / MI^b	M$_w$	M$_n$	M$_w$/M$_n$	DRI^c
CGC PE-1	0.908/1.0	109300	49450	2.2	1.1
CGC PE-2	0.908/1.0	90300	40860	2.2	4.0
CGC PE-3	0.908/1.0	89400	38500	2.3	14.0
LLDPE	0.924/1.0	119600	31300	3.8	N/A
mPE	0.911/1.2	118400	51500	2.3	0

a - Density, g/cm^3; b - Melt index, g/10 min; c - Dow Rheology Index

Table 2. Molecular parameters of HDPE blends

Resin	MI	M$_w$	M$_n$	M$_w$/M$_n$
HDPE-1, A(100%*)	0.31	254800	13200	19.3
B(80%)	0.70	205000	14300	14.3
C(60%)	1.53	152500	14100	10.8
D(40%)	3.32	139600	15000	9.3
E(20%)	7.76	112700	6200	6.9
HDPE-2, F(0%)	16.0	74000	6100	4.6

*composition ratio of HDPE-1 in blends

Table 3. Capillary and orifice die used

Diameter, mm	L/D ratio
0.254 (0.01")	0.5, 5, 10, 40
0.381 (0.015")	0.5, 20
0.761 (0.03")	0.5, 20

EXPERIMENTAL

RESIN

Four types of polymer were used in this study: CGC polyethylene and metallocene polyethylene (mPE), linear low density polyethylene (LLDPE) and high density polyethylene (HDPE). Each polymer has unique characteristics, depending on the polymerization process and catalyst system used. The CGC polyethylenes have various levels of LCB, as indicated by the value of the Dow Rheology Index (DRI). The metallocene polyethylene, which has a DRI of zero, has no LCB. Two types of HDPE were blended in order to vary the MW and the MWD in a systematic way, since it is practically impossible to vary the MWD while holding the MW constant. HDPE blends were prepared using a Berstorff (Model ZE25) corotating twin-screw extruder. Tables 1 & 2 show the molecular parameters of all samples.

EXPERIMENTAL PROCEDURE

Most experiments were carried out using a variable-speed, piston-driven Instron capillary rheometer, using a standard barrel diameter of 0.925cm. The onset of gross melt fracture was determined by visual inspection of the shape of the extrudate. The pressure drop and flow rate at the OGMF were recorded using both orifice and capillary dies (Table 3). All the dies have a 90° total entrance angle. The orifice dies were fabricated with a modified exit geometry to prevent the extrudate from touching the die exit. The entrance pressure drop data were determined using an orifice die. The true wall shear stress in the capillary die was determined by use of the Bagley end correction, while the true wall shear rate was determined using the Rabinowitch correction. All experiments were carried out over a broad temperature range, which included the extrusion temperature used in industry.

The Rheometrics Dynamic Analyzer (RDA-II) was used to determine the linear viscoelastic properties: storage modulus (G'), loss modulus (G") and complex viscosity (η^*). Samples were prepared by compression molding. Parallel plate fixtures (25 mm diameter) were used with a gap of 1.0 mm and experiments were conducted under a nitrogen atmosphere to prevent degradation. The frequency was varied in the range of 0.0126-500 (rad./sec). Strain sweeps were carried out to determine the linear viscoelastic regime for each material at the various test temperatures and frequencies. To estimate the parameters of power-law viscosity model, a regression technique was used. The power law exponent, n, is especially important as it is used as an index for estimating the extensional stress in Binding's and Cogswell's approximate analyses. All experiments were carried out at least three times to ensure reproducibility of results. Finally, critical extensional stress and strain rate at OGMF were estimated by using Cogswell's and Binding's analyses for the free convergence flow, even though all dies used have a 90° total entrance angle.

RESULTS AND DISCUSSION

Figure 1 shows the flow curve obtained for HDPE-1 using capillary and orifice dies. The onset of gross melt fracture is indicated by the arrows. While oscillatory flow was observed in the capillary dies before OGMF, this did not appear when the orifice dies were used. This implies that oscillatory flow originates in the die land. It is believed that slip-stick and melt compressibility are essential elements of oscillatory flow in the flow of conventional linear polymers. All samples used in this study show behavior similar to HDPE-1. Figure 2 shows the effect of the degree of LCB on the viscosity of CGC and metallocene polyethylenes. It is clear that the levels of LCB in CGC polyethylenes enhance the zero shear viscosity dramatically. Figure 3 shows the effect of high and low molecular weight fractions on complex vis-

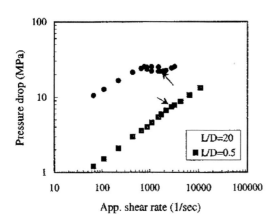

Figure 1. Flow curve for HDPE-1 (diameter = 0.015 inch, temperature = 180°C.

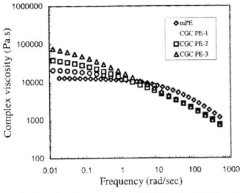

Figure 2. Complex viscosity curve of CGC and mPE (R = 2.5 cm, gap = 1.0 mm, temperature = 150°C).

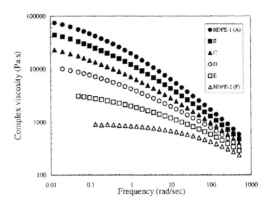

Figure 3. Complex viscosity curve of HDPE blends (R = 2.5 cm, gap = 1.0 mm, temperature = 160°C).

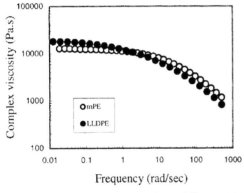

Figure 4. Complex viscosity curve of two different MWD samples (R = 2.5 cm, gap = 1.0 mm, temperature = 150°C).

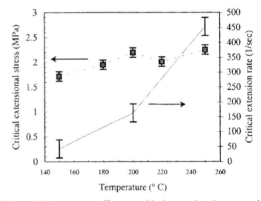

Figure 5. Temperature effect on critical extensional stress and strain rate (CGC PE-1, Cogswell's analysis).

cosity. Decreasing the high molecular weight fraction decreases the zero shear viscosity, as expected. However its contribution to shear thinning is not so strong. The effects of MWD on the complex viscosity are shown Figure 4. LLDPE has a broad MWD and shows shear thinning behavior at lower frequencies than mPE.

Figure 5 shows the effect of temperature on the critical extensional stress and strain rate. The critical extensional stress is almost independent of temperature over a wide range of temperatures. In contrast, the critical extensional strain rate shows a very strong temperature dependence. This implies that the critical extensional stress is an intrinsic property of a material. We therefore adapted the critical extensional stress as a criterion for OGMF. Figure 6 shows the effect of molecular weight on the critical extensional stress and strain rate as esti-

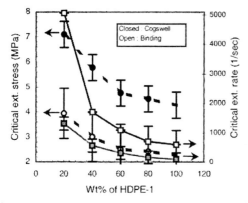

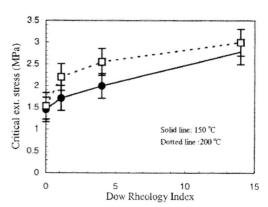

Figure 6. MW effect on critical extensional stress & strain rate (Cogswell's and Bindings's analyses, 160°C).

Figure 7. LCB effect on critical extensional stress and strain rate (Cogswell's analysis).

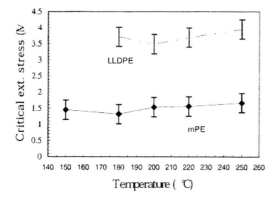

Figure 8. Comparison of critical extensional stress between LLDPE and mPE (Cogswell's analysis).

mated by Cogswell's and Binding's analyses. It shows that both the critical extensional strain rate and critical extensional stress are very sensitive to the low molecular weight fraction. In both cases, the values obtained using Binding's analysis are lower than those obtained using Cogswell's. This is related to the different assumptions of the two analyses.

The effect of degree of LCB is shown in Figure 7. Increasing the degree of LCB enhances the critical extensional stress. However, it seems that there is a critical level of LCB for enhancing the critical extensional stress. Finally, the effect MWD on the critical extensional stress is shown in Figure 8. LLDPE, which has a broad MWD without LCB, has a higher critical extensional stress than mPE, which has a narrow MWD and no LCB.

CONCLUSIONS

The critical extensional stress and critical strain rate at OGMF were determined using the entrance pressure drop analyses of Cogswell and Binding, and we propose the critical extensional stress as the appropriate criterion. The critical extensional stress was shown to be independent of temperature over a wide range of temperature. This is in contrast with the critical extensional strain rate, which is strongly temperature dependent. The proposed criterion for OGMF was evaluated using CGC and metallocene polyethylene and conventional linear low and high density polyethylenes. In CGC polyethylenes, the existence of very low degrees of long chain branching enhances the critical extensional stress as compared to purely linear polyethylenes. High density polyethylene blends having high amounts of low molecular weight fractions displayed higher critical extensional stresses than blends with lower amounts of low molecular weight material. Finally, polymers having a broad molecular weight distribution exhibit higher critical extensional stresses than those with narrow molecular weight distributions.

ACKNOWLEDGMENTS

This work was supported by the Natural Sciences and Engineering Research Council of Canada. The authors wish to thank Nova Chemicals and The Dow Chemical company for supplying the resins used and for size exclusion chromatography (SEC) results. The SEC measurements of the HDPE samples were carried out at the Samsung General Chemical research center in Taejon, Korea.

REFERENCES

1   F. N. Cogswell: *Polym. Eng. Sci.*, **12**, p.64 (1972).
2   D. M.Binding: **Techniques in Rheological Measurement**, Chapter. 1, *Chapman & Hall* (1993).
3   J. M. Dealy, K. F. Wissbrun: **Melt Rheology and its Role in Plastics Processing**, *Van Nostrand Reinhold* (1990).
4   C. S. Petrie, M. M. Denn: *AIChE J*, **22**, p.209 (1976).
5   S. A.White, A. D. Gotsis, D. G.Baird: *J. Non-Newt. Fluid Mech.*, **24**, p. 121 (1987).
6   M. M. Denn, G. Marrucci: AIChE Journal, 17, No.1, p.101 (1971).
7   F. N. Cogswell: Applied Polym. Symposium, 27, p.118(1975).
8   A. Ghijsels, C. H. C. Massardier, R.M.Bradley: Int. Polym. Proc., XII, p. 147 (1997).
9   A. E. Everage, R. C. Ballman: *J. Applied Polym. Sci.*, **18**, p.933 (1974).
10 M. T. Shaw: *J. Applied Polym. Sci.*, **19**, p. 2811 (1975).
11 H. N. Htirlimann, W. Knappe: *Rheol. Acta*, **11**, p.292 (1972).
12 A. G. Gibson: **Rheological Measurement**, Chap. 3, *Elsevier* (1988).
13 C. Kwag and J. Vlachopoulos: *Polym. Eng. Sci.*, **31**, p.1015 (1991).
14 M. Padmanabhan and CW. Macosko: *Rheol. Acta*, **36**, p.144-151 (1997).
15 M. E. Mackay, G. Astarita: *J. Non-Newt. Fluid Mech.*, **70**, p.219-235 (1997).

Relationship Between Structure and Rheology of Constrained Geometry Catalyzed and Metallocene Polyethylenes

Paula Wood-Adams and John M. Dealy
Department of Chemical Engineering, McGill University, Montreal, Canada

INTRODUCTION

Constrained geometry catalyzed and metallocene polyethylenes, referred to as mPEs, have narrow MWDs with polydispersity indexes of approximately 2. In the case of ethylene α-olefin copolymers produced with these catalysts, the SCBs are distributed randomly and. uniformly along the backbone and homogeneously among the molecules. Within the general class of mPE there are two subclasses: linear and branched mPEs. The linear mPEs, have either no branches or only short chain branches but no LCB. The branched mPEs, which are produced by INSITE[TM] Technology, have precisely controlled levels of LCB. These materials are referred to as substantially linear to distinguish them from highly branched LDPE.

The narrow MWD provides the good physical properties characteristic of mPE.[1] However, it also causes decreased shear thinning, resulting in higher energy requirements for processing. The LCB in the AFFINITY[TM] materials increase the amount of shear thinning increase melt strength and reduce susceptibility to melt fracture and draw resonance[2] and therefore improve their processability.

The LCB found in mPEs produced with constrained geometry catalysts is a very important molecular characteristic. One of the most reliable analytical methods for detecting and quantifying LCB is nuclear magnetic resonance (NMR).[3] However by using NMR alone it is difficult to distinguish between short branches that are 6 carbons in length or longer and long branches (which are in this case approximately the same length as the backbone). Therefore, another technique for the quantification of LCB is desired. Since rheological properties are very much affected by LCB they provide the basis for such a technique.

The primary objective of this work was to perform a thorough study of the effect of molecular structure on the rheology of mPEs with a focus on the effect of LCB. Because of the precise control of molecular structure that is possible with the constrained geometry catalysts we were able to study several sets of materials that differed in only one molecular characteristic. Therefore, we were able to study and describe independently the effects of average molecular weight, short chain branching, and long chain branching on the rheological behavior of these materials. A secondary objective was to develop a technique for quantifying LCB using rheological data.

EXPERIMENTAL MATERIALS AND TECHNIQUES

Three sets of mPEs were studied:
(1) three linear homopolymers with polydispersity indexes of 2 ± 0.1 and weight average molecular weights (M_w) ranging from 41900 to 359 000;
(2) three linear butene copolymers with polydispersities of 2.1 ± 0.02 and butene contents ranging from 1.4 wt% to 21.2 wt%; and
(3) five homopolymers with essentially the same MWDs but varying degrees of LCB (Table 1).

Table 1. Characteristics of series of branched homopolymers

Resin	Mw	Mw/Mn	LCB/10^{4}C	η_o, Pa.s	λ_d, s	a (eqn. 4)
L1*	100900	2.08	0	5800	0.34	0.06
B1	88400	1.98	0.12	11700	3.5	0.10
B2	96500	1.93	0.37	31600	7.8	N/A
B3	101500	1.99	0.42	56600	16	0.15
B4	90200	2.14	1.21	241140	345	N/A

*contains 1.4 wt% of butene comonomer

The dynamic linear viscoelastic (LVE) behavior of these materials was studied using a Rheometrics Dynamic Analyzer II (RDA II). Nonlinear shear measurements were performed using a sliding plate rheometer (SPR). Uniaxial elongational flow measurements were performed using a Rheometrics Melt Elongational rheometer (RME). All rheological data presented were measured at 150°C.

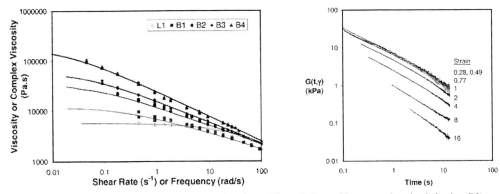

Figure 1. The effect of LCB on complex and shear viscosity.   Figure 2. Separable stress relaxation behavior (B3).

EFFECT OF MOLECULAR STRUCTURE ON THE RHEOLOGY OF mPEs

It was found that the linear mPEs follow the relation between zero shear viscosity (150°C) and M_w in Eqn 1. The coefficients in Eqn 1 are in good agreement with those reported by Raju *et al.*[4] for polyethylene.

$$\eta_o = 3.9 \times 10^{-15} M_w^{3.65} \qquad\qquad [1]$$

We found that SCB has no significant effect on the LVE behavior.

The complex viscosity curves for the high density mPEs are compared in Figure 1. The presence of LCB has four main effects on the complex viscosity:
   (1) the zero shear viscosity is increased for the same molecular weight,
   (2) the amount of shear thinning is increased,
   (3) the transition zone between the zero shear viscosity and the power law zone
      is broadened, and
   (4) the curve takes on a sigmoidal shape.

We also see the expected broadening of the relaxation spectrum with increased degree of LCB as demonstrated by the longest relaxation times (Table 1) calculated from the Doi-Edwards model (Eqn 2).

$$\lambda_d = \left(\frac{10}{\pi^2}\right)\eta_o J_s^o \qquad\qquad [2]$$

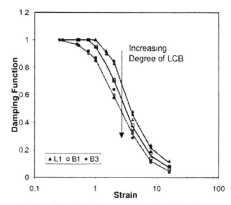

Figure 3. The effect of LCB on the damping function.

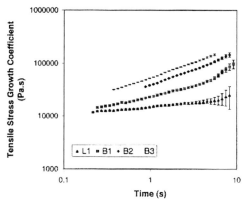

Figure 4. The effect of LCB on the tensile stress growth coefficient ($\dot{\varepsilon} = 0.5$).

It is interesting to note that the effect of increasing degree of LCB on the LVE behavior of mPEs is qualitatively similar to the effect of increasing the branch length of asymmetric stars that was seen by Gell *et al.*[5]

The viscosities of the branched materials are plotted in Figure 1 along with the complex viscosity. All of these materials obey the Cox-Merz rule, which is not the case with highly branched LDPE. This is an indication of the fact that we cannot extend accepted beliefs about the rheological behavior of LDPE to that of the branched mPE studied here.

Plotted in Figure 2 is an example of the stress relaxation behavior that was observed in the step strain experiments. These data can clearly be superposed by a vertical shift, meaning that the time and strain dependencies can be separated as in Eqn. 3.

$$G(t,\gamma) = h(\gamma)G(t) \qquad\qquad [3]$$

The damping functions for the three materials that have been subjected to these experiments to date are plotted in Figure 3. An increase in LCB results in the onset of nonlinear behavior at lower strains and increased damping at all strains. Again, this is not the case with LDPE, which exhibits less damping than linear polyethylene. It is often useful to fit an equation to the damping function. For these data we found that Eqn. 4 was adequate, and using more complicated equations did not significantly improve the fit. The fitted values for a are given in Table 1.

$$h(\gamma) = \frac{1}{1 + a\gamma^2} \qquad\qquad [4]$$

The tensile stress growth functions for several of the high density mPEs at a Hencky strain rate of 0.5 are plotted in Figure 4. As the degree of LCB increases, the tensile stress growth coefficient increases. We see also that the branched materials exhibit strain hardening while the curve for the linear material is essentially flat. It was not possible to obtain steady state values, i.e. the extensional viscosity, for the branched materials.

QUANTIFICATION OF LONG CHAIN BRANCHING

A useful rheological technique for inferring the level of LCB must meet three criteria:
   (1) It must distinguish between linear and branched mPEs
   (2) It must distinguish between branched mPEs and traditional polyethylenes
   (3) It must quantify LCB.
   To meet the first criterion, a technique must involve a parameter that is independent of M_w but highly sensitive to LCB. Since mPEs typically have polydispersity indexes of 2, the shape of the MWD is not a complicating factor in meeting this criterion. Traditional polyethylenes (LLDPE and HDPE) can have almost any shape and breadth of MWD, therefore to meet the second criterion the parameter must also be independent of MWD. Finally, to quantify LCB, the parameter must have a unique value for any given degree of LCB within a reasonable range.
   Initially extensional flow measurements were investigated as a basis for an LCB quantification technique. Since we were able to collect only transient data over relatively narrow strain range, and these data were quite noisy, we began to doubt their utility for this application. To establish precisely the information contained in the extensional data that was not present in the linear data, we made use of the reduced tensile stress growth coefficient defined in Eqn. 5.

$$\eta_{ER}^+(t,\dot{\varepsilon}) = \frac{\eta_E^+(t,\dot{\varepsilon})}{3\eta_o^+(t)} \qquad\qquad [5]$$

The linear shear stress growth coefficient (η_o^+) was calculated using the discrete spectrum. The reduced tensile stress growth coefficient curves are compared for four of the high density mPEs in Figure 5. We see that while the data for the linear material are distinct, the curves for the branched materials are not affected by the degree of LCB. Therefore, the extensional flow data do not contain any more information about the degree of LCB than is

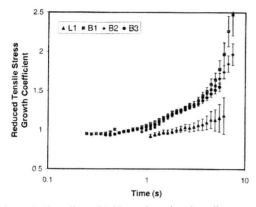

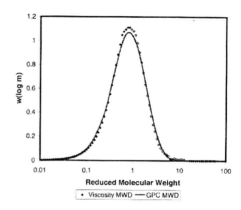

Figure 5. The effect of LCB on the reduced tensile stress growth coefficient ($\dot{\varepsilon} = 0.5$).

Figure 6. Comparison of viscosity and GPC MWDs for L1. Open symbols represent experimental data. Filled symbols represent extrapolated data.

contained in the LVE data. Since the LVE data are easier to measure and more precise than the extensional flow data we decided that they were more likely to be useful for inferring the level of long chain branching.

Since both LCB and MWD affect the LVE behavior, the challenge was to find a way to separate these two effects. If the material is known to be linear, then a reliable estimate of the MWD can be obtained using the LVE data.

Many techniques have been proposed to do this; a particularly useful one is that developed by Shaw and Tuminello[6] that allows the transformation of the complex viscosity curve into the MWD. For branched materials such a transformation is not possible, since the complex viscosity curve is strongly affected by degree of LCB. In fact for a branched material the MWD predicted from LVE data (referred to as the viscosity MWD) will deviate from the true MWD with the degree of deviation being related to the degree of LCB. Therefore, we decided to look for a technique for quantifying LCB based on comparing the gel permeation chromatography MWD with the viscosity MWD.

The technique used for calculating the viscosity MWD is a modified version of that developed by Shaw and Tuminello.[6] The transform equation is given below.

$$w(\log m) = \left[\frac{-\ln(10)}{m}\right]\left[\frac{\eta^*}{\eta_o}\right]^{1/\alpha}\left[\alpha\frac{d^2\ln\eta^*}{d\ln\omega^2} + \frac{d\ln\eta^*}{d\ln\omega} + \left(\frac{d\ln\eta^*}{d\ln\omega}\right)^2\right] \qquad [6]$$

where: $\omega = \omega_c m^{-\alpha}$ and $m \equiv M/\overline{M}_w$

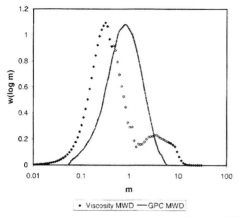

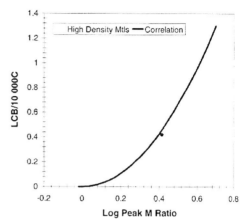

Figure 7. Comparison of viscosity and GPC MWDs for B3.   Figure 8. Peak molecular weight ration correlation.
Open symbols represent experimental data. Filled symbols
represent extrapolated data.

To use Eqn. 6, one must have η^* data that include the Newtonian plateau and the power law region. Since this range of behavior is not usually experimentally available, the experimental data must be extrapolated. We use the discrete relaxation spectrum to extrapolate at the low frequency end and the Vinogradov fluidity model to extrapolate at the high frequency end.

The accuracy of the viscosity MWD is illustrated by the data for the linear material, L1, in Figure 6. For this material and most other linear materials we have excellent agreement between the viscosity and GPC MWDs. A study was performed with several linear mPEs and LLDPEs to determine the effect of molecular weight, polydispersity, and short chain branching on the accuracy of the viscosity MWD. It was found that lowering the average molecular weight or increasing the polydispersity decreased the quality of the predicted MWD and that short chain branching had no effect on the quality. Even in the cases with the worst predictions for MWD, significant details of the distribution such as peak molecular weight were still well predicted.

The effect of LCB on the viscosity MWD can be seen in Figure 7, where the viscosity and GPC MWDs for material B3 are compared. The viscosity MWD is significantly different from the true GPC; the primary peak is shifted to a lower molecular weight, and a false peak is added at the high molecular weight end. The viscosity MWD can be thought of as the MWD of a linear polyethylene that would have the same complex viscosity curve as the branched, narrow MWD material. It was found that the location of the primary peak in the viscosity MWD relative to the peak in the GPC MWD was dependent on LCB. As the degree of LCB increases the primary peak shifts to a lower molecular weight. We were able to correlate the shifting of the primary peak to degree of LCB using Eqn. 7. The quality of this fit is shown in Figure 8.

$$\frac{LCB}{10^4 C} = 0 \text{ for } r_p < 1 \text{ and } \frac{LCB}{10^4 C} = 266\left[\log(r_p)\right]^2 \text{ for } r_p \geq 1 \qquad [7]$$

where: rp = GPC peak m/viscosity peak m.

CONCLUSIONS

The presence and degree of LCB significantly affects the linear and the nonlinear viscoelastic behavior of constrained geometry catalyzed polyethylenes. The rheological data most suitable for quantification of LCB are linear viscoelastic data. It was found that these data could be used in combination with gel permeation chromatography MWD data to predict the degree of LCB.

ACKNOWLEDGMENTS

This work was supported by the Natural Sciences and Engineering Council of Canada and The Dow Chemical Company. The authors wish to thank The Dow Chemical Company for supplying the resins, performing the molecular structure analyses (GPC and NMR) and allowing us to use their extensional rheometer.

NOMENCLATURE

G	relaxation modulus	η_E^+	tensile stress growth coefficient
h	damping function	η_{ER}^+	reduced tensile stress growth coefficient
J_s°	steady state compliance	η_o	zero shear viscosity
M_w	weight average molecular weight	η^*	complex viscosity
$\dot{\varepsilon}$	Hencky strain rate	ω	frequency
γ	strain	ω_c	critical frequency
λ_d	longest relaxation time		

REFERENCES

1   Lai, S. and G. W. Knight, SPE-ANTEC, 1188-1192 (1993).
2   Lai, S. T. A. Plumley, T. I. Butler, G. W. Knight, C. I. Kao, SPE-ANTEC, 1814-1815 (1994).
3   Randall, J.C., Am. Chem. Soc. Symp., 142, 93-118 (1980).
4   Raju, V.R., G.G. Smith, G. Marin, J.R. Knox, W.W. Graessley, *J. Polym. Sci.: Phys. Ed.*, **17**, 1183-1195 (1979).
5   Gell, C. B., W. W. Graessley, V. Efstratiadis, M. Pitsikalis, N. Hadjichristidis, *J. Polym. Sci. B; Polym. Phys.*, Vol **35**, 1943-1954 (1997).
6   Shaw, M. T. and W. H. Tuminello, *Polym. Eng. Sci.*, **34**, 159-165 (1994).

The Influence of Sequence Length Distribution on the Linear Crystal Growth of Ethylene-Octene Copolymers

John E. Wagner, Samir Abu-Iqyas, Kenneth Monar and Paul J. Phillips
Dept of Materials Science and Engineering. University of Tennessee, Knoxville, TN 37996-2200

INTRODUCTION

Linear low-density polyethylenes (LLDPE) are a commercially important class of polyethylenes, produced through the copolymerization of ethylene and comonomers such as hexene or octene, thereby producing butyl or hexyl branches, respectively. In order to separate the effects of copolymer content from molecular weight, a series of cross-fractionated copolymers has been investigated and reported in the recent past.[1,2] The copolymers which had been cross-fractionated for us by Dow Chemical had been synthesized using traditional Ziegler-Natta type catalysts, which are well-known to produce molecules in which the copolymer content varies with molecular weight. What was not well known at that time, or even now, was the sequence length distributions of the octene and ethylene mers within each molecule. It has been assumed to be random in the absence of contradictory information.

The regime theory which describes linear crystal growth in flexible polymers comprises two separate processes. The first process is the deposition of secondary nuclei on the growth face, usually denoted as occurring at a rate, i, the second process being the subsequent growth along the face at the niches formed by the secondary nuclei, often referred to as the rate of surface spreading, and denoted by the rate, g. The relative rates of these two processes determine the regime at which the crystallization occurs. The concept of transitions was first introduced in 1972 by Lauritzen and Hoffman[2] and has since been extensively evaluated by Phillips and others.[3-6] Studies of the cross-fractionated Ziegler-Natta ethylene-octene copolymers showed clearly that the effect of copolymerization was to reduce the linear growth rate. When analyzed further it was demonstrated that copolymerization reduced the rate of secondary nucleation in an exponential manner, whilst reducing the rate of surface spreading only in a somewhat linear manner. The net result was a depression of the regime I - regime II and regime II - regime III transitions to lower temperatures by a few degrees. The effect of increasing molecular weight was to negate the effect of increasing copolymer content as far as the transition temperature was concerned, but also to reduce the linear growth rate.

Similar copolymers have now been produced using metallocene catalysts which are believed to produce random copolymers because of the nature of catalytic process. Dow Chemical have now synthesized for us metallocene copolymers with molecular characteristics as near as possible identical to those of the cross-fractionated samples. Prior studies of the melting point - lamellar thickness relations have shown that the equilibrium melting points are depressed by a factor greater than that predicted by the Flory equation. It will be shown in this paper that for one of these copolymers the linear growth is depressed much more that in the equivalent fraction of a Ziegler-Natta polymer, the regime I - regime II transition temperature being depressed much more and regime III appears. The conclusion resulting is that the fractions of the Ziegler-Natta polymer contained non-random inclusion of the octene mers and that this study generates a very important new variable for crystallization studies, namely the sequence length distribution.

EXPERIMENTAL

LLDPE fractions, copolymers synthesized using metallocene catalysts and linear fractions were supplied and characterized by the Dow Chemical Co. All copolymers contain octene as the comonomer.

The fractions have been described previously.[1,2] In sample designation, L and H refer to low molecular weight and high molecular weight, respectively. The suffix M refers to materials synthesized using metallocene catalysts and ZN to materials synthesized using Ziegler-Natta catalysts (the latter are all fractions). Samples denoted by LPE are the linear samples which contain no branching. Table 1 gives the molecular weight characteristics of the samples.

Linear spherulitic growth kinetics experiments were performed using an Olympus polarizing microscope with an attached 35 mm camera and temperature controlled hot stage. The change in the morphological size was measured by taking photographs as a function of time. Samples were held at the melt tempera-

Table 1. Molecular weight characteristics

Sample	M_n	M_w	Branches*
LPE-13/18	13040	18120	0
LPE-54/101	53900	101300	0
L4-ZN	13400	23600	4.22
L4-M	27300	59900	3.98
H7-M	43600	94000	6.84
L11-M	21200	43700	10.86

* branches/1000 CH_3

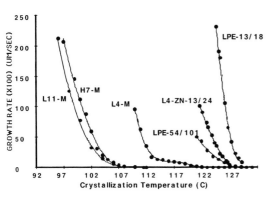

Figure 1. Linear growth rates of the samples as a function of crystallization temperature.

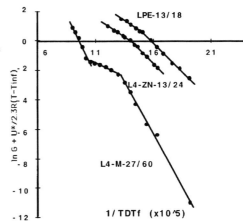

Figure 2. Secondary nucleation plots of linear growth rates showing the effect of catalyst type.

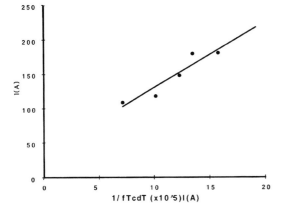

Figure 3. Relation between lamellar thickness (l) and inverse supercooling for L4-M.

ture of 150°C for 5 minutes and then rapidly quenched to the crystallization temperature. The linear growth rate data for L4-ZN and LPE-13/18 was taken from the work of Lambert et al.[1] and Hoffman et al.,[7] respectively.

RESULTS AND DISCUSSION

The variation of linear growth rate with temperature is shown in Figure 1 for four different materials. The additional linear fraction with higher molecular weight was studied so as to give some indication of the influence of the higher molecular weight of the metallocene polymer on the results. The crystallization temperature is reduced at 60×10^{-3} μm/s from approximately 126 to 106°C from LPE13/18 to L4-M. The results for the linear fraction confirm the well-known effect of reduction of growth rate by increasing molecular weight. The metallocene copolymer L4-M has had its growth rate depressed significantly compared to the LLDPE fraction L4-ZN. Observation of the data for the copolymers studied show that there are three regions to the plot of crystallization rate versus temperature curve. The gradual appearance of the regions is obviously a function of number of branches. L4-M shows all three regions, Figure 2, where two changes of

Table 2. Equilibrium melting point values

Sample	T_m^o, $^\circ$C^1	T_m^o, $^\circ$C^9
LPE-13/18	142.4	
LPE-54/101		142.7
L4-ZN	142.3	
L4-M		139.3
H7-M		140.4
L11-M		134.9

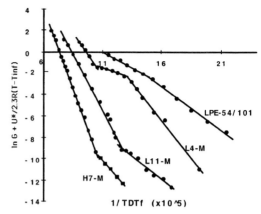

Figure 4. Secondary nucleation plots of linear growth rates showing the effect of increasing branching.

slope can be clearly discerned prior to any data manipulation. The effect is somewhat complicated by the fact that the molecular weight of the metallocene copolymer is about double that of the fraction. Man-Ho Kim[9] studied the affect of crystallization temperatures on lamellar thickness (l) for L4-M, Figure 3. The linear relationship between (l) and (T_c) is strong evidence that the different slopes in the regime plots are not a result of phase changes but because of the changes of i and g values with T_c. In order to analyze the data further and to attempt to further define the differences it is necessary to carry out conventional secondary nucleation analysis using:

$$G=G_0 \exp (U^*/R(T-T_\infty)) \exp [jb_0\sigma\sigma_0T_\infty]/fT\Delta T\Delta H_f] \qquad [1]$$

where j=4 for regimes I and III and j=2 for regime II.

The values for T_∞ used in this study are shown in Table 2. The values of LPE-54/101 and L4-M were determined from small angle scattering studies and Gibbs-Thompson extrapolation.[8] The values for LPE-13/18 and L4-ZN were those used previously.[1-3] Figure 2 shows that crystallization occurs for the linear molecules LPE-13/18 and for L4-ZN only in regime I and regime II. This analysis transforms the change of slope in the crystallization - temperature curve into clear changes of slope. Indeed, using the conventional interpretations of such behavior in polyethylene it can be said that crystallization occurs for the metallocene copolymer L4-M in regimes I, II, and III, and for the metallocene copolymers L11-M and H7-M in regimes II and III (Figure 4). These conclusions are reached also for reasonable variations of all parameters, including the T_∞ which is the most critical of all the parameters. Since it is already well-established that fitting of polyethylene data is virtually independent of variations in the parameters of the mobility exponential, a fact that is true here also, it appears that the differences between L4-ZN and L4-M are clear and established by the analysis. As one pro-

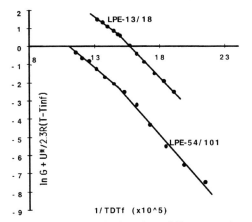

Figure 5. Secondary nucleation plots of linear growth rates showing the effect of molecular weight.

gresses from right to left in Figure 2 the effect of the catalysts used in synthesis can be seen. The Ziegler-Natta catalysts have higher crystallization temperatures and higher growth rates while the metallocene catalysts have lower crystallization temperatures and lower growth rates. Figure 5 shows the effect of molecular weight. LPE-13/18 with a lower molecular weight than LPE-54/101 has the higher crystallization temperature and higher growth rate but both show regimes I and II. In Figure 4 it is demonstrated that increasing branching is more important than molecular weight. As can be seen for the series LPE-54/101, L4-M, and L11-M as well as LPE-54/101, L4-M, and H7-M higher branching lowers the growth rate value offsetting the differences in molecular weight.

Hoffman and Miller[7] have demonstrated the use of secondary nucleation theory and experimental results to estimate the rate of deposition of secondary nuclei, i, and the rate of surface spreading, g, found in regime theory. These parameters can only be estimated at the regime transition temperature. The equations used for this study are too detailed to list here but are dealt with thoroughly by Hoffman and Miller and had been used previously by Lambert and Phillips.[1,2] The results obtained are presented in Table 3, where the values obtained by Lambert and Phillips have been combined with those currently calculated. All data is for the regime I - II transition point except for L11-M and H7-M which has data reported for the

Table 3. Regime transition analysis results

Sample	Transition temperature, °C	Rate cm/s	Growth rate x10^{-3} cm/s	Reeling-in x10^6 ncl/cm	g x10^{-5} cm^{-1}
LPE-13/18 (I-II)	125.3	1.73x10^{-4}	-	458	5.2
LPE-54/101 (I-II)	125.6	9.46x10^{-6}	1.83	560	4.64
L4-ZN-13/24 (I-II)	124.2	4.47x10^{-5}	-	13.7	4.73
L4-M (I-II)	119.5	5.45x10^{-6}	1.64	109	4.12
L4-M (II-III)	113.5	1.52x10^{-5}	1.44	1850	3.63
H7-M (II-III)	115.1	4.28x10^{-9}	1.49	1410	3.78
L11-M (II-III)	114.2	5.00x10^{-9}	1.50	2980	3.85

regime II-III transition point and L4-M which shows both the regime I-II transition point and the regime II-III transition points. The most striking result of the tabulated data is the remarkably constant value of the rate of surface spreading. Of course, there is the expected reduction between LPE-3/18 and LPE-54/101, see Figure 5, expected because of the increase in molecular weight and presumably a result of a decrease in the rate of reptation. L4-M and L4-ZN copolymers have approximately the same rate of surface spreading, close to the value of LPE-54/101. However, L4-M, L11-M, and H7-M show a small reduction in g values, where regimes II and III appear.

It is phenomenologically clear that the actual rate of surface spreading, or, by inference, the rate of reptation, is the controlling variable. The regime transition temperature occurs at whatever temperature is necessary for the rate of secondary nucleation to equal the rate of spreading, when expressed in equivalent units, of course. The molecular variables alter the rate of secondary nucleation in an exponential manner and the I-II regime transition occurs whenever that rate equals the scarcely varying rate of surface spreading. From the point of view of traditional nucleation theory the rate of spreading is essentially a rate of tertiary nucleation, which should be sensitive to molecular variables in a similar, but not identical, manner to secondary nucleation. Clearly this is not happening, a fact which has been alluded to before.[11] It therefore appears that the rate of surface spreading is essentially controlled by a local mobility rate, i.e., the rate of reptation. Perhaps this should not be a surprise as it was shown by Hoffman and coworkers a decade ago that the rate of reptation is just large enough to permit the measured value of the rate of surface spreading in linear polyethylene at the regime I - II transition. At lower crystallization temperature the rate of secondary nucleation increases.

The second important conclusion lies in the interpretation of the major difference between the LLDPE fraction (L4-ZN) and the metallocene copolymer (L4-M) of identical comonomer content. It is clear from previous work on the effect of molecular weight on the crystallization behavior of fractions, and from the results on the molecular weight dependence of linear polymers in the present study, that the influence of the molecular weight disparity is minimal and that the differences, which are major, have to be ascribed to another variable. The only remaining variable is the distribution of the sequence lengths of methylenes in the polymers. Metallocene copolymers are widely regarded as random copolymers and there is no evidence for non-randomness in the distribution from NMR studies. This aspect of their behavior is ascribed to the catalyst in the synthesis being present in solution. They are often referred to as single site catalysts. Ziegler-Natta copolymers are well-known to have complex copolymer content - molecular weight characteristics because of the catalysts being present on the surfaces of, and in the pores of, substrate particles. This gives rise to the widely known variety of copolymer content present in any material. When combined with the likelihood of chain scission occurring in the vicinity of an incorporated monomer (tertiary carbon at the branch) the well-known characteristics of high molecular weight - high methylene sequence length and low molecular weight - low methylene sequence length emerge. These characteristics are, of course, average characteristics over the materials produced during fractionation experiments. It was for this reason that the fractions studied[1,2] were produced by cross- fractionation, in which the materials were first fractionated by copolymer content. The state of micro-chemical analysis does not permit the estimation of the sequence length distribution in a copolymer. The most that can be produced is evidence regarding short sequences of the comonomer present in the polymers. The differences in composition of LLDPE molecules are determined by the surface characteristics of the substrate particles, by the relative abilities of the monomers to penetrate the pores in which the catalyst sites are located and by local chemical interactions. It is also believed that catalyst sites can become poisoned during polymerization, changing their characteristics. It seems likely therefore that the sequence length distribution of the copolymer molecules need not be ideally random and, indeed, might be expected to be non-random.

In the course of isothermal crystallization it is to be expected that the longest polymethylene sequence lengths will crystallize first. This principle has long been recognized and was first enunciated mathematically in Andrews' copolymer crystallization theory.[11] Similarly in dynamic crystallization such long polymethylene sequences will be expected to crystallize at higher temperatures than shorter sequences. Hence, a copolymer having an unusually large fraction of long polymethylene sequences will be expected to behave in a manner that is more similar to linear polyethylene than a copolymer containing an average number of long polymethylene sequences, predictable using random copolymerization theory. We therefore suggest that the differences in behavior between these two copolymers of the same octene content be ascribed to differences in polymethylene sequence distribution caused by the different polymerization methods.

It is suggested that the behavior reported here is the first firm evidence available for the inclusion of sequence length distribution in the list of essential molecular variables to be considered in understanding the behavior of copolymers, whereas all currently available theories consider only random copolymers.

CONCLUSIONS

Studies of a random ethylene-octene copolymer containing four octenes per 1,000 carbon atoms show that it behaves differently from a fraction of a Ziegler-Natta copolymer of the same octene content.

The random copolymer shows three different ranges of radial growth rate in the temperature range studied. When analyzed using secondary nucleation theory it is found to show all three regimes of growth. The range of temperature over which growth occurs is wider than that of both linear polyethylene and the aforementioned ZN fraction, but overlaps their ranges only in their low supercooling regions. They both show only regimes I and II. Increasing branch content shifts regimes from the I-II transition to II-III transition and reduces the growth rate significantly.

Analysis of the data yielded estimates of the rates of secondary nucleation and of surface spreading. The rate of surface spreading at the regime I-II transition and regime II-III transition scarcely varied across the polymers studied, whereas the rate of secondary nucleation varied significantly.

Differences between the random metallocene copolymer and the ZN fraction are ascribed to differences in the polyethylene sequence length distribution in the two types of copolymer synthesis.

ACKNOWLEDGMENTS

This research has been supported by the National Science Foundation under grants DMR9408678 and DMR9711986, by the Dow Chemical Company and by the Center of Excellence in Materials Processing at the University of Tennessee.

REFERENCES

1   Lambert, W. S.; Phillips, P. J., *Macromolecules*, 1994, **27**, 3537.
2   Hoffman, J. D.; Frolen, L. J.; Rose, G. S.; Lauritzan, J. I., *Res. Nat. Bur. Stand., Sect. A*, 1975, **79A**, 671.
3   Phillips, P. J., *Polym. Prepr.* (Am. Chem. Soc., Div. Polym. Chem), 1979, **20**, 438.
4   Hoffman, J. D., *Polymer*, 1983, **24**, 3.
5   Hoffman, J. D. and Miller, R. L., *Polymer*, 1997, **38**, 3151.
6   Phillips, P. J.; Lambert, W. S., *Macromolecules*, 1990, **23**, 2075.
7   Hoffman, J. D.; Miller, R. L., *Macromolecules* 1988, **21**, 3038.
8   Flory, P. J., *J. Chem. Phys.*, 1949, **17**, 223.
9   Kim, Man-Ho, 1996,"The Melting Behavior and Structure of Ethylene Copolymers from Metallocene Catalysts.", Ph.D. Dissertation, University of Tennessee.
10   Flory, P. J.; Vrij, A., *J. Am. Chem. Soc.*, 1963, **85**, 3548.
11   Andrews, E. H. Owen, P. J., and Singh A., *Proc. R. Soc. A*, 1971, **324**, 79.

Crystallization and Microstructure of Ziegler-Natta and Metallocene Based Isotactic Polypropylenes: Simulation and Experiment

Y. Churdpunt and A. I. Isayev

Institute of Polymer Engineering, The University of Akron, Akron, Ohio 44325-0301

INTRODUCTION

Metallocene catalysts seem to get great attention from both researchers and manufacturers. However, there are only few research papers concerning the crystallization of metallocene based i-PP.[1-3] Typically, the cooling rates obtained in crystallization experiments are low in comparison with those in the real material processing such as injection molding. Therefore, many researchers have used various methods to get the cooling rate as close as possible to the real process.[4-6] The quenching of slabs with a thermocouple placed in them is a very useful method to study the crystallization at high cooling rates. Using this technique, De Carvalho *et al.*[6] were able to modify the approach to the simulation of the crystallization process in quenched slabs of i-PP's.[5] They introduced the modified equation to obtain the nuclei concentration along the gapwise direction and incorporated the temperature dependence of physical properties. Their equation for calculating the nuclei concentration led to a better description of the spherulite diameters. In particular, simulated spherulite size distribution in quenched slabs was found to be in good agreement with the corresponding experimental results. In the present paper, quenching experiments and simulations of the morphology of metallocene and Ziegler-Natta based i-PP were performed along with DSC experiments to obtain isothermal and nonisothermal rate of crystallization and induction time. Also, spherulitic growth rate experiments were carried out. The differences and similarities of crystallization processes in these polymers were discussed.

Table 1. Fundamental properties of i-PP's

Polymer	MFR	Polydispersity	M_w	T_m, ^oC	$\cong H$, J/g	Ultimate crystallinity, %
mPP3904	22	2.0	161000	147.5	98.23	0.47
mPP3825	32	1.9	144800	148.7	105.66	0.51
ZNPP444	73	4.4	144000	162.0	121.90	0.58

EXPERIMENTAL

MATERIALS AND METHODS OF INVESTIGATION

Two commercial grades of metallocene based i-PP's, Achieve grades 3904 and 3825, supplied by Exxon Chemical Company and one grade of Ziegler-Natta based i-PP, H444, supplied by Himont USA, Inc. were used. The fundamental properties of these polymers are listed in Table 1. The symbol mPP3904 and mPP3825 represents metallocene based i-PP's grade 3904 and 3825, respectively, and ZNPP444 represents Ziegler-Natta based i-PP's grade H444. It is seen from Table 1 that mPP3825 has the same molecular weight as ZNPP444 sample. Isothermal and nonisothermal crystallization studies were performed by using a Perkin Elmer Differential Scanning Calorimeter (DSC-7). The evolution of heat during crystallization obtained from nonisothermal crystallization experiment at various cooling rates was used to calculate the degree of crystallinity as a function of temperature after the temperature lag correction. The master curve approach to crystallization kinetics was performed to obtain the nonisothermal rate constant from the nonisothermal experiment.[7] The nonisothermal rate constants from the nonisothermal experiments were combined with the nonisothermal rate constants determined from isothermal experiments to get the wide range of crystallization temperatures. The growth rate was obtained by measuring the diameter of spherulites as a function of time during isothermal and nonisothermal crystallization using a special device built in our laboratory.[5] Quenching experiments were carried out on sheets approximately 3.3 mm thick, 2H. An iron-constantan thermocouple connected to a PC with a data acquisition system was inserted into the sample in order to record the temperature trace during crystallization. The sample was heated to 200°C for about 10 minutes to remove any potential nuclei and quenched into various quenching mediums, i.e. water and air. The diameter of spherulites was measured across the slab using a Leitz 12 Pol S Optical microscope.

THEORETICAL CONSIDERATIONS

CRYSTALLIZATION PROCESS

The overall crystallization kinetics for isothermal crystallization has been widely described by the Avrami-Kolmogoroff equation.[8,9] The half time of crystallization, $t_{1/2}$, which is the elapsed time from the crystallization onset time until the relative crystallinity reaches a value of 0.5, has the form:

$$t_{1/2} = \left(\frac{\ln 2}{k} \right)^{1/n} \tag{1}$$

where k(T) is the isothermal rate constant as a function of temperature, T, and n is the Avrami exponent.

During polymer processing, crystallization takes place under the nonisothermal conditions. In order to describe the crystallization occurring in the real process, the nonisothermal crystallization model has been used. One of the most accepted equations was proposed by Nakamura et al.[10] The differential form of the Nakamura equation is

$$\frac{d\theta}{dt} = nK(T)(1-\theta)[-\ln(1-\theta)]^{(n-1)/n} \tag{2}$$

where K(T) is the non-isothermal crystallization rate constant which can be related to k(T) by

$$K(T) = [k(T)]^{1/n} = \frac{(\ln 2)^{1/n}}{t_{1/2}} \tag{3}$$

Based on Equation (3) and according to Hoffman et al.[11] the nonisothermal rate constant can be expressed as

$$K(T) = (\ln 2)^{1/n} \left(\frac{1}{t_{1/2}} \right)_o \exp\left(-\frac{U*/R}{T - T_\infty} \right) \exp\left(-\frac{k_k}{T \Delta T f} \right) \tag{4}$$

where R is the universal gas constant, T_m^o is the equilibrium melting point, $\Delta T = T_m^o - T$, and $f = 2T/(T + T_m^o)$.

From the expression, the kinetic model has four parameters: $(1/t_{1/2})_o$ is a pre-exponential factor that includes all terms independent of temperature, k_k is the nucleation exponent, U^* is the activation energy for the transport of crystalline units across the phase boundary and T_∞ is taken as the glass-transition temperature minus 30°C. The isothermal induction time, t_i, for the crystallization process was obtained from the nonisothermal experiments as described by Isayev *et al.*[12] The nonisothermal induction time, t_i, can be obtained by[13]

$$\bar{t} = \int_0^{t_i} \frac{dt}{t_i(T)} = 1 \qquad [5]$$

where $\bar{t}$ is the induction time index which becomes unity when the quiescent crystallization starts. The isothermal induction time can be assumed to follow the Godovsky and Slonimsky[14] equation

$$t_i = t_m(T_m^o - T)^{-a} \qquad [6]$$

where t_m and a are material constants independent of temperature.

Growth rates, $G(T)$, can be determined from Hoffman-Lauritzen growth rate equation[11] as follows

$$G(T) = G_o \exp\left(-\frac{U^*/R}{T - T_\infty}\right) \exp\left(-\frac{k_g}{T\Delta Tf}\right) \qquad [7]$$

where G_o is a pre-exponential factor independent of temperature. The remaining factors have the same meaning as outlined in Equation 4.

To calculate the gapwise distribution of spherulite diameter, the nuclei concentration, $N(T)$, can be determined by

$$N(T) = \frac{3V_\infty k(T)}{4\pi G(T)^3} \qquad [8]$$

where V_∞ is the maximum volume fraction of spherulites at infinite time. The spherulite diameter is assumed to be equal to L which is the average distance between nuclei. According to Equation 8, for instantaneous three-dimensional growth, the constant L takes the form

$$L = \left(\frac{V_\infty}{8\pi \overline{N}} \right)^{1/2}$$ [9]

where $\overline{N}$ is the average nuclei concentration. Considering $\overline{N}$ as an average for the whole crystallization process based on the remaining amorphous fraction, because new nuclei can be formed only in this phase, the average nuclei concentration can be expressed as follows:

$$\overline{N} = \int_0^1 N(T)(1-\theta)d\theta$$ [10]

with N(T) given by Equation 8.

SIMULATION OF MICROSTRUCTURE AND CRYSTALLIZATION IN QUENCHED SLABS

In order to predict the crystallinity distributions and morphology of i-PP slabs of finite thickness, an infinitely extended slab is considered. For this case, a one dimensional heat transfer, incorporating a heat of crystallization term, was used. The rate of crystallization was obtained from the Nakamura model.

RESULT AND DISCUSSION

QUIESCENT CRYSTALLIZATION KINETIC PARAMETERS

The result for the ultimate degree of crystallinity is listed in Table 1. Ziegler-Natta based i-PP shows the highest value of the ultimate degree of crystallinity and melting point compared to metallocene based i-PP's which is also found by Bond and Spruiell.[3] Moreover, the higher molecular weight of metallocene based i-PP's shows higher ultimate degree of crystallinity.

Figures 1 and 2 show the nonisothermal rate of crystallization and the spherulitic growth rate, respectively. The nonisothermal rate of crystallization of metallocene based i-PP's increased as the molecular weight increased. When comparison was made at the same molecular weight between metallocene and Ziegler-Natta based i-PP's, metallocene based i-PP has a lower crystallization rate constant. The growth rates of both grades of metallocene based i-PP's are the same but lower than the Ziegler-Natta based i-PP. However, metallocene based i-PP's have a much lower melting point compared to the conventional Ziegler-Natta based i-PP as indicated in Table 1. This makes the difference in the degree of supercooling, which plays an important role as the driving force of crystallization and should be taken into consideration. Bond and Spruiell[3] suggested that the lower melting point of metallocene based i-PP's is caused by the less stable lamellae and is due to the defects being more uni-

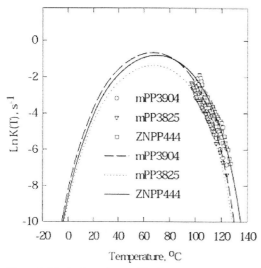

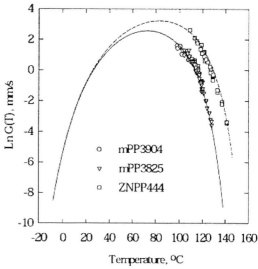

Figure 1. Nonisothermal crystallization rate constants as a function of temperatures for various i-PP's. Symbols represent experimental data while lines represent non-linear fit to the Hoffman-Lauritzen equation (Equation 4).

Figure 2. Spherulite growth rates as a function of temperatures for various i-PP's. Symbols represent experimental data while lines represent non-linear fit to the Hoffman-Lauritzen growth equation (Equation 7).

formly distributed from chain to chain compared to the Ziegler-Natta based i-PP. The fitted curves for the growth rate and the nonisothermal rate constant were obtained by a nonlinear fit to the kinetic equation, Equation 4, and the growth equation, Equation 7. The values of the parameters obtained from the fittings are required for the simulation of spherulite size. From our experiences, the predicted spherulite sizes are very sensitive to these values. Very high cooling rates or at the low temperatures can not be achieved in our DSC and growth rate experiments. Therefore there is still an uncertainty for the fitting at the low temperature region, even though an attempt was made to incorporate the isothermal and nonisothermal experiments to extend the range of the temperatures.[7,15,16] Unless this problem is solved, it may introduce some errors in the prediction of spherulite size. This low temperature region is more important than the high one because it is the-temperature region where crystallization in the real process occurs, especially in the quenching process in this study.

INDUCTION TIME PARAMETERS

In order to simulate the temperature profile, the induction time parameters, a and t_m, in Equation 6, were obtained by fitting the experimental induction time from DSC and quenching experiments as a function of cooling rates for each polymer. Figure 3 shows the temperature traces with the measured and predicted induction time of slabs during quenching in air. The

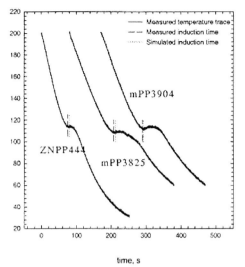

Figure 3. Temperature traces with the measured and predicted induction times during quenching of i-PP slabs in air at 25°C. Vertical dashed lines represent the experimental results while dotted lines represent the simulated data for mPP3904: 2H = 3.10 mm, h = 23 W/(m²K), position of thermocouple y/H = 0.13; mPP3825: 2H = 3.10 mm, h = 23 W/(m²K), position of thermnocouple y/H = 10; ZNPP444: 2H = 2.96 mm, h = 35 W/m²K, position of thermocouple y/H = 0.01.

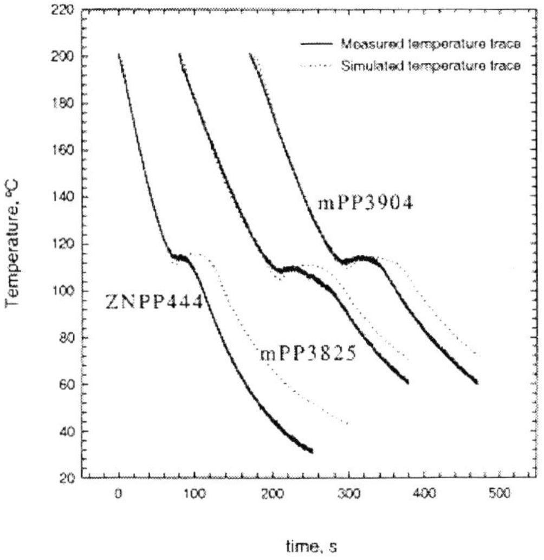

Figure 4. Measured (solid lines) and simulated (dotted lines) temperature traces during quenching of i-PP's slabs in air at 25°C for mPP3904: 2H = 3.10 mm, h = 23 W/(m²K), position of thermocouple y/H = 0.13; mPP3825: 2H = 3.10 mm, h = 23 W/(m²K), position of thermocouple y/H = 0.10; ZNPP444: 2H = 2.96 mm, h = 35 W/(m²K), position of thermocouple y/H = 0.01.

measured induction time is determined based on the times where the slope of the temperature profile is significantly changed.

SIMULATION OF THE TEMPERATURE TRACES

By using the induction time parameters and the best fit of heat transfer coefficient to the experimental temperature profile, we can simulate the temperature trace measured by the thermocouple. Figure 4 shows the measured and simulated temperature profiles during quenching in air for all polymers. The simulated temperature profile is fitted very well in the early stage of quenching and starts showing the discrepancy after the crystallization completed. It indicates that the heat transfer coefficient, h, being constant before and during crystallization is possibly not constant for the whole cooling process. However, because this discrepancy occurs after the completion of crystallization process, it should not affect the prediction of the spherulite size.

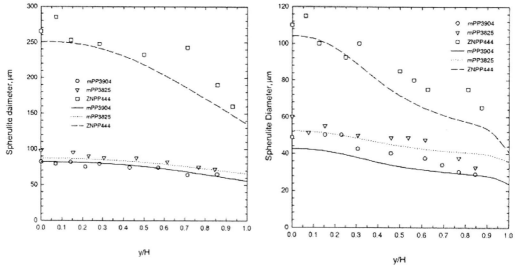

Figure 5. The gapwise distribution of measured (symbols) and simulated (lines) spherulite diameters of i-PP slabs quenched in air at 25°C.

Figure 6. The gapwise distribution of measured (symbols) and simulated (lines) spherulite diameters of i-PP slabs quenched in water at 25°C.

SPHERULITE SIZE DISTRIBUTION IN THE QUENCHED SLABS

The morphology developed in the quenched slab of all polymers has the spherulitic structure. The samples quenched in water usually show the spherulite diameters smaller than the samples quenched in air. Also, water quenched samples show a lot of very fine and bright spherulites at the surface. This type of spherulites were not observed in the samples quenched in air. Instead, the sample quenched in air showed large row nucleated spherulites at the surface.

As shown in Figures 5 and 6, Ziegler-Natta based iPP has much larger spherulites than metallocene based iPP's. Between two metallocene based i-PP's, the one with the lower molecular weight showed somewhat larger spherulite size. Moreover, samples quenched in air and water show an increase in size of spherulites as the distance from the wall increases. The distribution of the spherulite size is more evident in the samples quenched in water because the cooling rate is higher and much different from the wall to the core. It is obvious that all simulated results coincide with the experiment rather well. Simulated spherulite diameter distribution of the samples quenched in water showed more dramatic variation in sizes across the thickness than that in the air. This simulated result is in very good agreement with the morphology observed in the experiment.

CONCLUSIONS

Ziegler-Natta and metallocene based i-PP's exhibit different growth rates and rates of crystallization. Both experimental and simulated results from quenching showed that Ziegler-Natta based i-PP has much larger spherulite sizes than both grades of metallocene based i-PP's. Moreover, the lower molecular weight of metallocene based i-PP's produces the larger spherulite sizes.

REFERENCES

1   E. B. Bond and J. E. Spruiell, SPE ANTEC Tech. Papers, 388, 1754 (1997).
2   C. Y. Cheng and J. W. C. Kuo, SPE ANTEC Tech. Papers, 302, 1942 (1997).
3   E. B. Bond and J. E. Spruiell, SPE ANTEC Tech. Papers, 388, 1750 (1997).
4   Z. Ding and J. E. Spruiell, *J. Polym. Sci. Polym. Phys., Part B*, **35**, 1077 (1997)
5   A. I. Isayev and B .F. Catignani, *Polym. Eng. Sci.*, 37 (1997)
6   B. De Carvalho, R. E. S. Bretas and A. I. Isayev, *J. Appl. Polym. Sci.* (submitted).
7   T. Chan, L. Guo and A. I. Isayev, *SPE Tech Papers*, 41,1471(1995)
8   M. Avrami, *J. Chem. Phys.*, **7**, 1103 (1939); **8**, 212 (1940) ; **9**, 177 (1941)
9   A. N. Kolmogoroff, *Isvestiya Akad. Nauk SSSR, Ser.Math.*, **1**, 355 (1937)
10   K. Nakamura, K. Katayama and T. Amano, *J. Appl. Polym. Sci.*, **17**, 1031 (1975).
11   J. D. Hoffman, G. W. Davis and S.I. Lauritzen, **Treatise on Solid State Chemistry**, Vol.3, Ch7, N. B. Hannay ed., *Plenum*, New York (1976).
12   A. I. Isayev, T. W. Chan, K. Shimojo and M. Gmerek, *J. Appl. Polym. Sci.*, 55, 807 (1995).
13   W. L. Sifleet, N. Dinos and J. R. Collier, *Polym. Eng. Sci.*, **13**, 10 (1973).
14   Y. K. Godovsky and G. L. Slonimsky, *J. Polym. Sci. Polym. Phys. Ed.*, **12**, 1053 (1974).
15   T. W. Chan and A. I. Isayev, *Polym. Eng. Sci.*, **34**, 461 (1994).
16   T. W. Chan, G. D. Shyu and A. I. Isayev, *Polym. Eng. Sci.*, **35**, 733 (1995).

Crystallization of Ethylene-Octene Copolymers at High Cooling Rates

Paul J. Phillips and John Wagner

Department of Materials Science & Engineering University of Tennessee, Knoxville, TN 37996-2000

INTRODUCTION

Linear low-density polyethylenes (LLDPE) are a commercially important class of polyethylenes, produced through the copolymerization of ethylene and comonomers such as hexene or octene, thereby producing butyl or hexyl branches, respectively. In order to separate the effects of copolymer content from molecular weight, a series of cross-fractionated copolymers has been investigated and reported in the recent past.[1,2] The copolymers which had been cross-fractionated for us by Dow Chemical had been synthesized using traditional Ziegler-Natta type catalysts, which are well-known to produce molecules in which the copolymer content varies with molecular weight. What was not well known at that time, or even now, was the sequence length distributions of the octene and ethylene mers within each molecule. It has been assumed to be random in the absence of contradictory information. Our studies of random copolymers produced using metallocene catalysts showed that the crystallization behavior was very different from that of the Ziegler-Natta materials, a result which had to be a consequence of the ZN polymers being non-random within each molecule.

Quiescent crystallization is usually separated into its component parts of primary nucleation, linear spherulite (or lamellar) growth and secondary crystallization. In this paper only the linear spherulite growth rates will be considered. Regime theory, which describes linear spherulite growth in flexible polymers, comprises two separate processes. The first process is the deposition of secondary nuclei on the growth face, usually denoted as occurring at a rate i, the second process being the subsequent growth along the face at the niches formed by the secondary nuclei, often referred to as the rate of surface spreading, and denoted by the rate g. The relative rates of these two processes determines the regime at which the crystallization occurs. The concept of transitions was first introduced in 1972 by Lauritzen and Hoffman[2] and has since been extensively evaluated by Phillips and others.[3,4,5,6] Studies of the cross-fractionated Ziegler-Natta ethylene-octene copolymers showed clearly that the effect of copolymerization was to reduce the linear growth rate. When analyzed further it was demonstrated that copolymerization reduced the rate of secondary nucleation in an exponential manner, whilst reducing the rate of surface spreading only in a somewhat linear manner. The net result was a depression of the regime I - regime II and regime II - regime III transitions to lower temperatures by a few degrees. The effect of increasing molecular weight was to negate the effect of increasing copolymer content as far as the transition temperature was concerned, but also to reduce the linear growth rate. Up until the present date, it has been generally assumed that regime theory can describe all known crystallization behavior in polyolefins.

Similar copolymers have been produced using metallocene catalysts, which are believed to produce random copolymers because of the nature of catalytic process. Dow Chemical synthesized for us metallocene copolymers with molecular characteristics as near as possible identical to those of the cross-fractionated samples. Prior studies of the melting point - lamellar thickness relations have shown that the equilibrium melting points are depressed by a factor greater than that predicted by the Flory equation. It was shown in the ANTEC 98 paper that for one of these copolymers the linear growth is depressed much more that in the equivalent fraction of a Ziegler-Natta polymer, the regime I - regime II transition temperature being depressed much more and regime III appears. The conclusion resulting was that the fractions of the Ziegler-Natta polymer contained non-random inclusion of the octene mers and that this study generated a very important new variable for crystallization studies, namely the sequence length distribution.

In the study being reported in ANTEC 99, the studies of the metallocene copolymers have been extended to the very high supercoolings characteristic of commercial processing operations. It will be demonstrated that, under such conditions, the polymer behavior changes dramatically and no longer is dependent on comonomer content and, apparently also, molecular weight. This process is envisioned as being similar to a congealing process in which the copolymer molecules wiggle into place as the growing crystal sweeps through them, the hexyl branches being included in the metastable crystals. Presumably, these crystals anneal and perfect through a variety of subsequent processes, in which some of the hexyl branches are ejected from the crystals. It is suggested that these later events control the final mechanical and physical properties, rather than the effective crystallization temperature.

EXPERIMENTAL

LLDPE fractions, copolymers synthesized using metallocene catalysts and linear fractions were supplied and characterized by the Dow Chemical Co. All copolymers contain octene as the comonomer. The fractions have been described previously,[1,2] as also have the metallocene

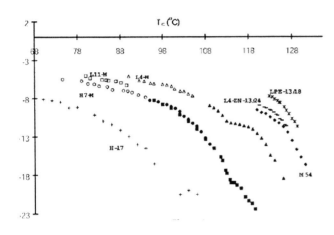

Figure 1. Growth rate.

copolymers.[7] In sample designation L and H refer to low molecular weight and high molecular weight respectively. The suffix M refers to materials synthesized using metallocene catalysts and ZN to materials synthesized using Ziegler-Natta catalysts (the latter are all fractions). The numbers following the letters refer to the number of hexyl branches per 1,000 carbon atoms. Samples denoted by LPE are the linear samples, the numbers following the LPE referring to the molecular weights in 1,000s.

Linear spherulitic growth kinetics experiments were performed using an Olympus polarizing microscope with an attached 35 mm camera and temperature controlled hot stage. The change in the morphological size was measured by taking photographs as a function of time. Samples were held at the melt temperature of 150°C for 5 minutes and then rapidly quenched to the crystallization temperature. Crystallization experiments at high supercoolings were carried out in the original Ding - Spruiell rapid cooling equipment.[8,9]

RESULTS AND DISCUSSION

The variation of linear growth rate with temperature is shown in Figure 1 for several different materials, where the filled symbols represent points obtained in conventional isothermal crystallization experiments. The open symbols represent data points obtained in rapid cooling experiments, where the polymer generates its own pseudo-isothermal crystallization temperature. It can be seen clearly that as high supercoolings are approached the curves of all the copolymers are tending to merge into a single curve, regardless of comonomer content or molecular weight. It has to be remembered that the equilibrium melting point is dependent on molecular weight and comonomer content and the data should be corrected for that variable using the supercooling, relative to the equilibrium melting point of each copolymer. When the data are plotted as a function of supercooling (Figure 2), the same merging phenomenon can be seen, perhaps even more clearly. The third type of plot normally made is the secondary nucleation, or regime, plot in which the logarithm of the growth rate, less a mobility term, is plotted against the reciprocal of supercooling. It is this plot that permits the different regimes

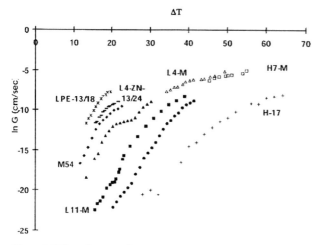

Figure 2. Effect of supercooling.

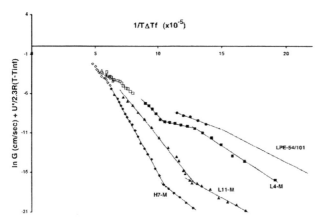

Figure 3. Crystallization behavior.

to be identified and is, in many ways, the defining plot for the crystallization behavior of a polymer. Such a plot is shown in Figure 3.

The entire crystallization behavior of the copolymers can now be analyzed. First, the filled points, indicative of conventional isothermal studies, show the expected decrease in crystallization rate as comonomer content is increased (e.g. compare L4 with L11). Also apparent is the decrease of crystallization rate with increasing molecular weight (e.g. H7 is slower than L11). Also shown are the data for the Ziegler-Natta fraction, (L4-ZN), which crystallizes much more rapidly than for the comparable metallocene copolymer L4-M. When the open symbols, indicative of rapid cooling experiments, are considered, it can be seen that initially they continue on the same lines as the filled points. This is a very important observation, since it tells us that there is no significant difference between an experiment carried out isothermally and one that is carried out dynamically in a rapid manner, and in which the polymer generates its own pseudo-isothermal crystallization condition. It is, however, clear that as the experiments proceed to successively higher and higher cooling rates that there is a slow downward change in the slope for all the copolymers. Ultimately, the copolymers are attaining the same growth rates. This cannot occur if the normal selection rules of copolymer crystallization, based on exclusion of the comonomer units, continue to apply. Nor can the merging of the curves to a common line, regardless of molecular weight, occur unless there is essentially a lack of long range movement of the polymer chains during the crystallization process. In other words, the copolymer molecules, essentially, must freeze into place without the exclusion of molecular defects. This is a new form of crystallization, quite different from the one occurring at higher temperatures, where the molecules have time to reject the comonomer species and to diffuse in a way which is dependent on their molecular weights.

It can also be noticed that in the isothermal regions (filled points) that the slopes of the lines are increasing as the comonomer content increases. From the equations governing regime theory, this can only be a result of two possible changes. The first is an increase in the fold surface free energy, which appears in the numerator of the slope, probably caused by crowding of the rejected side branches in the interfacial regions. Secondly, there may be a reduction in the latent heat of fusion, which appears in the denominator, caused by incorporation of hexyl branches in the crystal. The equilibrium melting point of the copolymers will be decreased if there exists a substantial incorporation of defects, and so the equilibrium melting points of the copolymers may be progressively decreasing as crystallization temperature decreases. If this is the case, then our estimates of effective supercooling used in Figures 2 & 3 will be high. This in itself could be responsible for the decreasing slope found at very high supercoolings (open symbols). Additionally, if large amounts of hexyl branches are incorporated in the crystals, rather than being excluded, then the crystal will expand, lowering the latent heat of fusion. At the same time, the crowding in the interfacial regions will be reduced, thereby causing a reduction of the fold surface free energy. All of these effects would result in a decrease in the slope. The effects are likely to be occurring simultaneously. It should also be noted that expansion of the crystal lattice, through incorporation of defects, will reduce the surface free energy by allowing more surface area per emerging chain in the fold surfaces.

It would seem unlikely that the metastable crystals formed at the growth front would retain their initial structure and condition. They would be expected to try to reduce their free energy by annealing processes, which would result in the ejection of substantial numbers of hexyl branches to the fold surfaces, or even the lateral surfaces, of the crystals. Such processes could be occurring immediately after the crystals are formed, or perhaps later during storage at ambient temperature, or both. The stabilization processes could occur so rapidly that they would not be detected in normal characterization experiments, or they might require some form of thermal activation and hence could be studied independently. The type and character of the stabilization processes will be controlled to some extent by the detailed structure of the folded surface. Presumably, this crystallization process is very similar to the Flory switchboard model, there being many tie-molecules. Also the number of adjacent reentry folds will be very restricted. Under these conditions, there would be very little molecular mobility as the ability of the molecules to snake through the crystal would be very much inhibited. These restrictions would lead to very complex morphological structures on a micro-scale, as the crystals stabilize, involving the crystals, the interfacial regions and the interlamellar material. They would also lead to complex molecular trajectories and hence network structures. It would be these morphological - molecular features that would interact to control the resultant mechanical and physical properties. The ultimate properties, such as strain to failure and fracture toughness might be affected most by these features.

CONCLUSIONS

The random copolymers show up to three different ranges of radial growth rate in the isothermal temperature range studied, assigned to all three regimes of growth. Increasing branch content shifts regimes from the I-II transition to II-III transition and reduces the growth rate significantly.

When studies are extended to rapid cooling experiments, in which the polymer generates its own pseudo-isothermal crystallization temperature, the applicable existing regime III region is continued for a restricted range of temperature. When very high supercoolings are approached a new form of crystallization occurs, in which the comonomer units are no longer excluded and molecular weight dependencies appear to be eliminated. The copolymers become indistinguishable from one another on the basis of their spherulite growth rates.

ACKNOWLEDGMENTS

This research has been supported by the Polymers Program of the National Science Foundation under grants DMR9408678 and DMR9711986 and by the Center of Excellence in Materials Processing at the University of Tennessee.

REFERENCES

1   Lambert, W. S.; Phillips, P. J., *Macromolecules*, 1994, **27**, 3537.
2   Hoffman, J. D.; Frolen, L. J.; Ross, G. S.; Lauritzen, J. I., *J. Res. Nat. Bur. Stand., Sect. A*, 1975, **79A**, 671.
3   Phillips, P. J., *Polym. Prepr.* (Am. Chem. Soc., Div. Polym. Chem), 1979, **20**, 438.
4   Hoffman, J. D., *Polymer*, 1983, **24**, 3.
5   Hoffman, J. D. and Miller, R. L., *Polymer*, 1997, **38**, 3151.
6   Phillips, P. J.; Lambert, W. S., *Macromolecules*, 1990, **23**, 2075.
7   Kim. M-H; Phillips, P. J., *J. Appl. Polym.Sci.*, 1998, **70**, 1893.
8   Ding, Z.; Spruiell, J. E., *J. Polym. Sci., B, Polym. Phys. Ed.*, 1996, **34**, 2783.
9   Supaphol, P.; Sprueill, J. E., *J. Polym. Sci., B, Polym. Phys. Ed.*, 1998, **36**, 681.

Kinetics of Non-Isothermal Crystallization Syndiotactic Polypropylene: Avrami, Ozawa, and Kissinger Approaches

Pitt Supaphol and Joseph E. Spruiell
Dept. of Materials Science and Engineering, University of Tennessee, Knoxville, TN 37996

INTRODUCTION

Syndiotactic polypropylene (sPP) has largely been a laboratory curiosity since it was first produced in 1960s by Natta *et al.*[1,2] It has gained more interest in terms of industrial applications since 1988 when Ewen *et al.*[3] reported that highly stereoregular and regioregular sPP can be synthesized using novel metallocene catalysts. Since then, industrial applications of sPP have been extensively explored in the areas such as films,[4,5] injection molding,[6] and melt-spun fibers[7] by researchers at both Fina Oil and Chemical Company in Dallas, Texas and Mitsui Toatsu Chemical Company in Osaka, Japan, which are the business partners and major commercial producers of sPP. Other physical properties related to applications have also been investigated and reported.[8,9]

It is well known that a semi-crystalline polymer, such as sPP, will crystallize when subjected to a cooling regime whether it is isothermal or non-isothermal. The physical properties of semi-crystalline polymers are strongly dependent on the crystalline structure formed and the extent of the crystallization during processing. It is therefore very important to understand the processing-structure-property inter-relationships of the studied materials, which in this case is sPP. Studies related to the chain conformation, crystal structure, morphology, and phase transitions in sPP have been reported extensively in recent years. These researches up to 1995 were reviewed and published by Rodriguez-Amold *et al.*[10] Published studies on crystallization kinetics of sPP include isothermal Avrami crystallization kinetics,[11-14] and the isothermal kinetics of the linear growth rates.[11,15-17] There have been a number of published reports concerning non-isothermal crystallization behavior of sPP,[11,18-19] but none had reported on the kinetics of non-isothermal crystallization.

In this manuscript, we focus on the kinetics of nonisothermal crystallization of sPP using a differential scanning calorimeter (DSC). The data were analyzed based on the modified Avrami and Ozawa approaches. The activation energy for non-isothermal melt crystallization was also calculated based on the Kissinger method.

THEORETICAL

The most common approach used to describe the overall isothermal crystallization kinetics is the Avrami equation:[20]

$$1 - \frac{\chi_t}{\chi_\infty} = 1 - \theta(t) = \exp\left(-k_t t^n\right) \qquad [1]$$

where χ_t and χ_∞ are the transient and ultimate absolute crystallinity, respectively, $\theta(t)$ the relative crystallinity, k_t the isothermal bulk crystallization rate constant (which constitutes the growth and nucleation behavior), n the Avrami exponent, and t the time taken during the crystallization process (the incubation time, t_o, is excluded). Both k_t and n are constants typical of a given crystalline morphology, and type of nucleation at a particular crystallization condition.[21]

Alternatively, Equation (1) can be written in its logarithmic form:

$$\ln[-\ln(1 - \theta(t))] = \ln k_t + n \ln(t) \qquad [2]$$

Based on Equation (2), parameters k_t and n can readily be calculated from the least-square line fit to the double logarithmic plot of $\ln[-\ln(1-\theta(t))]$ versus $\ln(t)$, where k_t is taken as the anti-logarithmic value of the y-intercept and n is simply the slope (calculated for 10% to 80% relative crystallinity only).

In the study of non-isothermal crystallization using DSC, the energy released during the crystallization process appears to be a function of temperature rather than time as in the case of isothermal crystallization. Therefore, the relative crystallinity as a function of temperature, $\theta(T)$, can be formulated as

$$\theta(T) = \frac{\int_{T_o}^{T} \frac{dH_c}{dT} dT}{\Delta H_c} \qquad [3]$$

where T_o and T represent the onset and an arbitrary temperature, respectively, and dH_c the enthalpy of crystallization released during an infinitesimal temperature range dT, and ΔH_c the overall enthalpy of crystallization for a specific cooling condition.

In an attempt to use Equations (1) and (2) to analyze the non-isothermal crystallization kinetics data in DSC, we need to assume that the sample experiences the same thermal history as determined by the DSC furnace. This may be realized only when the thermal lag between the sample and the furnace is kept minimal. If this assumption is valid, the relation between crystallization time, t, and temperature, T, is given as

$$t = \frac{T_o - T}{\phi} \qquad [4]$$

where ϕ is the cooling rate. According to Equation (4), the horizontal temperature axis observed in a DSC thermogram can readily be transformed into the time scale.

Based on the mathematical derivation of Evans,[22] Ozawa[23] extended the Avrami theory[20] to be able to describe the non-isothermal case. Mathematically, the relative crystallinity can be written as a function of cooling rate as

$$1 - \frac{\chi_T}{\chi_\infty} = 1 - \theta(T) = \exp\left(-\frac{k_T}{\phi^m}\right) \qquad [5]$$

where χ_T is the absolute crystallinity developed at temperature T, χ_∞ the ultimate absolute crystallinity, k_T the rate function, and m the Ozawa exponent (which is similar to the Avrami exponent).

Analysis of the non-isothermal crystallization data can be performed through the use of the logarithmic form of Equation (5):

$$\ln[-\ln(1 - \theta(T)0] = \ln k_T - m \ln \phi \qquad [6]$$

By plotting $\ln[-\ln(1 - \theta(T))]$ versus $\ln(\phi)$ for a fixed temperature, the parameters k_T and m can be extracted as the anti-logarithmic value of the y-intercept and the negative value of the slope, respectively.

EXPERIMENTAL

The sPP sample used in this study was supplied in the pellet form by Fina Oil and Chemical Company in La Porte, Texas. Molecular characterization data, kindly measured by Dr. Roger

A. Phillips and his coworkers at Montell USA, Inc. in Elkton, Maryland, shows the following molecular weight information: M_n = 76 200, M_w = 165 000, M_z = 290 000, and M_w/M_n = 2.2. In addition, the syndiotacticity measured by ^{13}C NMR shows the racemic pentad content, [%rrrr], to be 77.1%, the racemic triad content, [%rr], to be 87.3%, and the racemic dyad content, [%r], to be 91.4%.

A film of sPP sample was prepared from sliced pellets melt-pressed at 190°C between a pair of Kapton films which in turn were sandwiched between a pair of stainless steel platens, in a Wabash compression molding machine under a pressure of 67 kpsi. After ten minutes holding time, the film of approximately 280 μm thick was taken out and immediately quenched in an ice-water bath, while it was still between the two steel platens. This treatment assumes that previous thermal and mechanical histories formed during the pelletization process were essentially erased, and provides a controlled starting condition for our experiments.

A Perkin-Elmer Series 7 Differential Scanning Calorimeter was used to study non-isothermal crystallization in this manuscript. The DSC7 equipped with internal cooling unit reliably provided a cooling rate up to 200°C min^{-1}. Temperature calibration was performed using an indium standard (T_m^o = 156.6°C and ΔH_f^o = 28.5 J g^{-1}). The consistency of the temperature calibration was checked every other run to ensure reliability of the acquired data. The DSC sample was in the form of disc cut from the film prepared. The sample weight was kept around 4.9 ± 0.3 mg.

The experiment started with heating the sample from 40°C at a scanning rate of 80°C min^{-1} to 190°C where it was held for 5 min before cooling down at a desired constant cooling rate, ϕ (in the range of 1°C min^{-1} to 100°C min^{-1}), to -40°C. The cooling exotherms were recorded and analyzed accordingly. It is worth noting that the 5 min holding time at 190°C is ample to erase the previous crystalline memory.

RESULTS AND DISCUSSION

NON-ISOTHERMAL CRYSTALLIZATION KINETICS BASED ON MODIFIED AVRAMI EQUATION

The crystallization exotherms of sPP for nonisothermal crystallization from the melt at 6 different cooling rates ranging from 1°C min^{-1} to 10°C min^{-1} are presented in Figure 1. It is evident that, as the cooling rate increases, the temperature at 1% relative crystallinity, $T_{0.01}$, the peak temperature, T_p, and the temperature at 99% relative crystallinity, $T_{0.99}$, are shifted to lower temperatures. It should be noted that $T_{0.01}$ and $T_{0.99}$ represent the temperatures at which the crystallization process began and ended, respectively. The values of $T_{0.01}$, T_p, and $T_{0.99}$, along with the corresponding enthalpy of crystallization, ΔH_c, are tabulated in Table 1.

Apparently, the values of $T_{0.01}$, T_p, and $T_{0.99}$ decrease as the cooling rate increases. This leads to the fact that the higher the cooling rate, the later the crystallization begins and ends

Table 1. Values of $T_{0.01}$, T_p, and $T_{0.99}$ and ΔH_c for non-isothermal crystallization exotherms of sPP

Cooling rate, °C min⁻¹	$T_{0.01}$, °C	T_p, °C	$T_{0.99}$, °C	ΔH_c, J g⁻¹
1	94.3	82.1	77.6	39.2
2	88.3	76.0	70.3	37.2
4	82.0	70.9	44.8	36.1
6	78.2	63.7	40.6	33.9
8	76.9	61.1	41.6	33.3
10	77.6	57.1	29.9	32.0

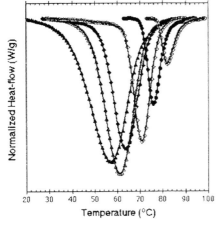 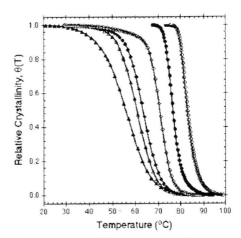

Figure 1. Non-isothermal melt crystallization exotherms of sPP at 6 different cooling rates (°C min⁻¹): O 1, ● 2, ◇ 4, ◆ 6, △ 8, ▲ 10.

Figure 2. Relative crystallinity as a function of temperature of sPP at 6 different cooling rates (°C min⁻¹): O 1, ● 2, ◇ 4, ◆ 6, △ 8, ▲ 10.

(based on the temperature domain). In addition, the enthalpy of crystallization also decreases as the cooling rate increases. This relates directly to the fact that the higher the cooling rate, the lower the ultimate absolute crystallinity obtained.

The relative crystallinity, $\theta(T)$, as a function of crystallization temperature for each cooling rate can be taken from the corresponding exotherm shown in Figure 1 using Equation (3), and is shown in Figure 2. As mentioned previously, in order for these data to be analyzed using the Avrami equation,[20] the horizontal temperature scale must be transformed into the time domain using the relationship defined in Equation (4). The plots of relative crystallinity, $\theta(t)$, as a function of time for various cooling rates are shown in Figure 3. It is apparent that the

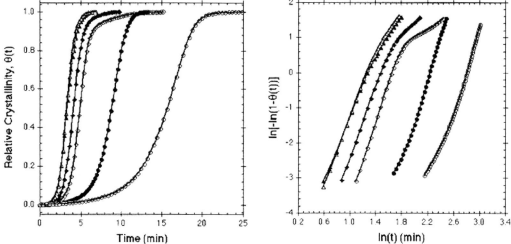

Figure 3. Relative crystallinity as a function of time of sPP at 6 different cooling rates ($^\circ$C min^{-1}): O 1, ● 2, ◇ 4, ◆ 6, △ 8, ▲ 10.

Figure 4. Typical Avrami analysis of the data shown in Figure 3 at 6 different cooling rates ($^\circ$C min^{-1}): O 1, ● 2, ◇ 4, ◆ 6, △ 8, ▲ 10.

Table 2. Non-isothermal crystallization kinetics of sPP based on modified Avrami approach

Cooling rate, $^\circ$C min^{-1}	n	k_t, min^{-n}	$t_{0.5}$, min	$t_{0.5}^{-1}$, min^{-1}
1	5.25	3.83×10^{-7}	15.64	0.06
2	6.08	1.23×10^{-6}	8.85	0.11
4	5.60	9.11×10^{-5}	4.89	0.20
6	5.00	5.87×10^{-4}	4.09	0.24
8	4.80	2.31×10^{-3}	3.26	0.31
10	4.48	3.10×10^{-3}	3.32	0.30

faster the cooling rate, the shorter the time needed for the completion of the crystallization process (though the plots for the cooling rates of 8°C min^{-1} and 10°C min^{-1} are very similar).

Based on Equation (1), Avrami crystallization kinetics parameters (n and k_t) can be extracted through the double logarithmic plots of $\ln[-\ln(1 - \theta(t))]$ versus $\ln(t)$, as mentioned previously. Such plots are shown in Figure 4. The values of n and k_t as well as the crystallization half-time, $t_{0.5}$ (taken directly from the data shown in Figure 3), and its reciprocal value, $t_{0.5}^{-1}$, are listed in Table 2.

The Avrami exponent, n, ranges from 4.5 to 6.1, and decreases in value as the cooling rate increases. The Avrami exponent observed suggests that the primary stage for

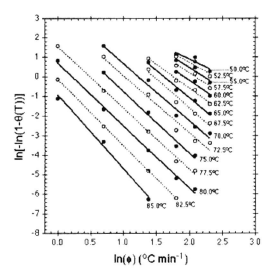

Figure 5. Typical Ozawa analysis of the data shown in Figure 2 at different temperatures.

non-isothermal melt crystallization may correspond to a three-dimensional sheaf-like growth with athermal nucleation at low cooling rates (from 1°C min⁻¹ to 6°C min⁻¹), and to a three-dimensional spherical growth with thermal nucleation at higher cooling rates (greater than 8°C min⁻¹).[21] More recently, anomalous observation of n (i.e., n > 3) can be explained systematically using the "nucleation rate theory," developed by Ding and Spruiell,[24] and it is currently under further investigation.

The rate of non-isothermal crystallization can readily be described by the values of the Avrami rate constant, k, and the crystallization half-time, $t_{0.5}$ (or more specifically, the reciprocal value of the crystallization half-time, $t_{0.5}^{-1}$). The result shows that the rate of non-isothermal crystallization is proportional to the cooling rate. In other words, sPP crystallizes faster as the cooling rate increases. However, for this particular sPP sample, the cooling rate may not be higher than 10°C min⁻¹ in order for the completion of the crystallization process to occur during cooling from the melt. At cooling rates greater than 10°C min⁻¹ some crystallizable material will still be uncrystallized as the temperature drops into the sub-glass transition region (the glass transition, T_g, of this sample was determined to be -6.1°C),[14] and it will crystallize upon subsequent heating (this process is known as cold crystallization).

NON-ISOTHERMAL CRYSTALLIZATION KINETICS BASED ON OZAWA EQUATION

By simply replacing t in Equation (1) with T/φ, Ozawa[23] was able to extend the Avrami equation to describe the kinetics of non-isothermal crystallization using a DSC. The data can be taken directly from the raw data shown in Figure 2, and the analysis can be done through the double logarithmic plot of $\ln[-\ln(1 - \theta(T))]$ versus $\ln(\phi)$. Figure 5 illustrates such plots for temperatures in the range of 50 to 85°C. A least-square line is drawn to data points taken for each temperature. The rate function, k_T, is taken as the anti-logarithmic value of the y-intercept, and the Ozawa (or Avrami) exponent, m, is taken as the negative value of the slope. Values of k_T and m as well as correlation coefficient, r^2, of the fit are listed in Table 3.

Table 3. Non-isothermal crystallization kinetics of sPP based on Ozawa approach

Temperature, °C	m	kT, (°C min^{-1})^m	r^2
50.0	1.66	6.81×10^1	0.85
52.5	1.88	8.82×10^1	0.90
55.0	2.19	1.27×10^2	0.92
57.5	2.23	1.03×10^2	0.95
60.0	2.33	8.65×10^1	0.96
62.5	2.48	7.69×10^1	0.99
65.0	2.82	9.43×10^1	0.99
67.5	3.00	7.90×10^1	0.97
70.0	2.94	3.79×10^1	0.98
72.5	2.93	1.94×10^1	0.97
75.0	3.18	1.15×10^1	0.99
77.5	3.13	4.51	1.00
80.0	3.20	2.00	1.00
82.5	3.37	8.47×10^{-1}	1.00
85.0	3.72	3.75×10^{-1}	0.99

Based on the plots shown in Figure 5 and the correlation coefficients, r^2, listed in Table 3, the Ozawa approach[23] is found to be a satisfactory description of the non-isothermal crystallization kinetics of sPP, especially in the temperature range of 57.5 to 85°C where the r^2 value is greater than 0.95. The Ozawa exponent, m, is found to increase with an increase in temperature, and lies in the range of 2.3 to 3.7 (within the temperature range of 60 to 85°C). The result is in a good agreement with our earlier results[14] on isothermal bulk crystallization kinetics of sPP which showed that the Avrami exponent, n, ranges from 2.7 to 3.2 (within the same temperature range).

ACTIVATION ENERGY FOR NON-ISOTHERMAL MELT CRYSTALLIZATION BY KISSINGER METHOD

According to the data listed in Table 1, it is obvious that the peak temperature, T$_p$, is a certain function of cooling rate. This relationship can be described by an Arrhenius-type equation. Such a relationship was introduced by Kissinger,[25] which can be written in the following form:

$$\ln\left(\frac{\phi}{T_p^2}\right) = c_1 - \frac{E_a}{R}\frac{1}{T_p}$$ [7]

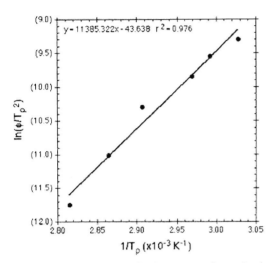

Figure 6. Estimation of activation energy for molecular transport based on non-isothermal melt crystallization using Kissinger method.

where c_1 is an arbitrary constant, and R the universal gas constant. E_a, is the activation energy required for transportation of molecular segments from the bulk which is in the molten state to the crystal growth surface. Based on Equation (7), the slope of the least-square line drawn through the plot of $\ln(\phi / T_p^2)$ versus $1/T_p$ equals the activation energy, $-E_a/R$, thus the activation energy, E_a, can be calculated accordingly (i.e., $E_a = -R \times$ slope). Such a plot is shown in Figure 6. With the correlation coefficient being 0.98, the bulk of the data complies well with the Kissinger equation. From the slope of the least-square line, the activation energy, E_a, for non-isothermal crystallization from the melt of this sPP sample is found to be -94.7 kJ mol^{-1}.

We have also calculated the activation energy, E_a, for non-isothermal melt crystallization from the raw data published in reference 11 (Table 5.2). Values of E_a, along with some important molecular characterization data of various sPP samples are listed in Table 4. E_a values of two different kinds of poly(aryl ether ketone)[27,28] are also listed in Table 4 for comparison.

Table 4. Activation energy, E_a, for non-isothermal melt crystallization of various sPP samples and other polymers

Polymer	M_w	M_n	[%rrrr]	E_a, kJ mol^{-1}	Ref.
sPP	165000	76200	77	-94.7	this work
	22900	20800	88	-97.3	26
	45900	41700	86	-78.0	26
	158400	132000	87	-86.1	26
PEEKK*		30000		-189	27
PEDEKmK*		10000		-219	28

*poly(aryl ether ketone) of different kinds

Obviously, the activation energy for non-isothermal crystallization from the melt of sPP samples are comparable, with the average value of -89.0±8.8 kJ mol^{-1}. As mentioned previously, the activation energy, E_a, relates to the ability of the molecule to extricate itself from the melt onto the crystal growth surface. In fact, this ability is controlled by a number of factors, one of which is molecular stiffness. This effect can be illustrated with the fact that the E_a values of poly(aryl ether ketone)s are much higher in magnitude than those of sPP samples.

The activation energy for molecular transport, E_d, is often calculated from the combination of the Arrhenius-type and WLF equations.[29] Eckstein *et al.*[30] reported the E_d value of a sPP sample (M_w = 137 600, M_n = 62 800, M_w/M_n = 2.2, and [%rrrr] = 79.6%) to be 73.7 kJ mol[-1]. Evidently, the values of E_a and E_d calculated for sPP are comparable and related. The discrepancy between the values obtained from these two different methods may be a result of the uncertainty in the peak temperature measured from a DSC non-isothermal crystallization exotherm. Thermal lag is likely the most probable cause for the unreliability in the peak temperature value obtained, hence the E_a value calculated.

CONCLUSIONS

The non-isothermal melt crystallization exotherms showed that as the cooling rate increases the temperature at 1% relative crystallinity, the peak temperature, and the temperature at 99% relative crystallinity moves towards the lower temperature range. The crystallization half-time is found to decrease in value as the cooling rate increases. This suggested that crystallization is completed in a shorter time period as the cooling rate increases. Analysis of the time dependent relative crystallinity revealed that the Avrami exponent lies in the range of 4.5 to 6.1. Whereas, analysis of the temperature dependent relative crystallinity suggested that the Ozawa exponent ranges from 1.7 to 3.7 (within the temperature range of 50 to 85°C).

Using the Kissinger method, the activation energy for non-isothermal melt crystallization was determined to be -94.7 kJ mol[-1]. This value was found to be in a good agreement with the values calculated from the nonisothermal crystallization data of other sPP samples available in the literature. Comparison of the value with the activation energy for molecular transport determined from the WLF equation suggested that these two parameters may be related.

REFERENCES

1   G. Natta, I. Pasquon, P. Corradini, M. Peraldo, M. Pegoraro, and A. Zambelli, *Rend. Acd. Naz. Lincei.*, **28**, 539 (1960).
2   G. Natta, I. Pasquon, and A. Zambelli, *Ibid.*, **84**, 1488 (1962).
3   J. A. Ewen, R. L. Johns, A. Razavi, and J. D. Ferrara, *J. Am. Chem. Soc.*, **110**, 6255 (1988).
4   J. Schardl, L. Sun, S. Kimura, and R. Sugimoto, Society of Plastics Engineers, Annual Technical Conference (ANTEC) Proceedings, 3414 (1995).
5   J. Schardl, L. Sun, S. Kimura, and R. Sugimoto, *J. Plastic Film & Sheeting*, **12**, 157 (1996).
6   L. Sun, E. Shamshoum, and G. DeKunder, Society of Plastics Engineers' Annual Technical Conference (ANTEC) Proceedings, 1965 (1996).
7   M. Gownder, Society of Plastics Engineers' Annual Technical Conference (ANTEC) Proceedings, 1511 (1998).

8   W.R. Wheat, Society of Plastics Engineers' Annual Technical Conference (ANTEC) Proceedings, 2275 (1995); 565 (1997).

9   K. Yoshino, X.H. Yin, R. Sugimoto, N. Uchikawa, M. Kawata, T. Murakami, and H. Kato, Institute of Electrical Engineers of Japan's 1995 Int'l Symp. on Electrical Insulating Materials Proceedings, 355 (1995).

10   J. Rodriguez-Arnold, Z. Bu, and S. Z. D. Cheng, *J. M. S.-Rev. Macromol. Chem. Phys.*, **C35**(1), 117 (1995).

11   J. Rodriguez-Arnold, in Ph.D. Dissertation, University of Akron, 1994.

12   J. Rodriguez-Arnold, A. Zhang, S. Z. D. Cheng, A. J. Lovinger, E. T. Hsieh, P. Chu, T. W. Johnson, K. G. Honnell, R. G. Geerts, S. J. Palackal, G. R. Hawley, and M. B. Welch, *Polymer*, **35**(9), 1884 (1994).

13   P. Supaphol, J. J. Hwu, P. J. Phillips, and J. E. Spruiell, Society of Plastics Engineers' Annual Technical Conference (ANTEC) Proceedings, 1759 (1997).

14   P. Supaphol and J. E. Spruiell, *submitted to J. Polym. Sci., Polym. Phys.*

15   R. L. Miller and E. G. Seeley, *J. Polym. Sci., Polym. Phys.*, **20**, 2297 (1982).

16   J. Rodriguez-Arnold, Z. Bu, S. Z. D. Cheng, E. T. Hsieh, T. W. Johnson, R. G. Geerts, S. J. Palackal, G. R. Hawley, and M. B. Welch, *Polymer*, **35**(24), 5194 (1994).

17   P. Supaphol and J. E. Spruiell, *submitted to Polymer.*

18   J. Sclunidtke, G. Strobl, and T. Thurn-Albrecht, *Macromolecules*, **30**, 5804 (1997).

19   G. Strobl, *Acta Polymer.*, **48**, 562 (1997).

20   M. Avrami, *J. Chem. Phys.*, **7**, 1103 (1939); **8**, 212 (1940); **9**, 177 (1941).

21   B. Wunderlich, in **Macromolecular Physics**, Vol. 2, *Academic Press,* New York, 1976, page 147.

22   U. R. Evans, *Trans. Faraday Soc.*, **41**, 365 (1945).

23   T. Ozawa, *Polymer*, **12**, 150 (1971).

24   Z. Ding and J. E. Spruiell, *J. Polym. Sci., Polym. Phys.*, **35**, 1077 (1997).

25   H. E. Kissinger, *J. Res. Nat'l. Bur. Stand.*, **57**, 217 (1956).

26   Calculated from the non-isothermal crystallization data listed in Table 5.2 of the reference 11.

27   T. Liu, Z. Mo, S. Wang, and H. Zhang, *Polym. Eng. Sci.*, **37**(3), 568 (1997).

28   T. Liu, Z. Mo, and H. Zhang, *J. Appl. Polym. Sci.*, **67**, 815 (1998).

29   D.W. Van Krevelen, in **Properties of Polymers,** Vol. 3, *Elsevier,* Amsterdam, 1990, pages 595-598.

30   A. Eckstein, C. Friedrich, A. Lobbrecht, R. Spitz, and R. Muelhaupt, *Acta Polymer.*, **48**, 41 (1997).

High Temperature Flexible Polyolefins a User's Perspective

Atul Khare, Samuel Y Ding, Michael T. K. Ling, and L. Woo
Baxter Healthcare, Round Lake, IL 60073

INTRODUCTION

MEDICAL MATERIAL REQUIREMENTS

Before we proceed, it would be rational to examine some of the special requirements specific to the medical device and packaging industries, that set our needs apart from other markets. Medical industry always attaches a premium to material's optical clarity. A material's ability to allow light transmission un-impeded is not just a desire to be "clean", but rather rooted in good clinical practice. For example, in many primary packaging, prior to administration of the medical fluid to the patient, the nurse is required to inspect visually for particulate matter (PM) contamination. This inspection is required to protect patients against the danger of exposure to the visible contaminants. In situations where medications in the powder form need to be mixed or compounded prior to administration, the ability to visually inspect for the completion of dissolution is another important safeguard for patient's safety. Recently, transfusion therapies involving harvested, stored, and cultured cellular components became widespread. During the in-vitro phase, these cellular components frequently need to be manipulated through separation, fractionation, and incubation for diagnostic or therapeutic purposes. In many of the steps, the ability to inspect contents visually or microscopically is crucial to the clinical protocol. These are just a few reasons optical clarity is a prerequisite criterion for material selection.

Another attribute specific to the medical industry is the need to deliver therapies in a sterile manner. As a consequence, compatibility with common sterilization methods is a must. These methods include: steam autoclaving, ethylene oxide and ionizing radiation (either gamma or electron beam). Ethylene oxide is currently being phased out due to worker exposure and environmental concerns, while long term shelf life stability is an issue for the material post radiation sterilization. Steam autoclaving has a long history in sterilization dating back to the canning preservation of food in the 19th century. Steam is benign, non-toxic, and effective, although it creates a serious challenge on the material properties with the high temperature (121°C) and pressure.

A third requirement that has received increasing attention is the environmental compatibility. As professionals concerned with healthcare, we obviously cannot, in the course of mending our patients, destroy the environment that we live in and our children will inherit. Of course, the quantitative "environmental score" is quite elusive and often subjected to passionate debate. However, some of the basic tenets are very obvious; progressive companies have already adopted many of the practices in this direction.

THERMOPLASTIC ELASTOMER PROPERTIES

In order to achieve thermoplastic elastomer (TPE) property, there must be thermally reversible physical crosslinks on the polymer chain. For the familiar styrene diene block copolymers of the ABA type, when the outer blocks are the glassy styrene domains, phase separation from incompatibility forces the styrene blocks to aggregate into domains that trap the elastomeric diene segments between them. At elevated temperatures above the styrene glass transition (T_g), the overall structure becomes deformable and shapeable, giving rise to the thermoplastic property. Therefore, a necessary condition for the thermoplasticity is the existence of phase separated blocky sequences on the main chain where soft, elastomeric blocks are flanked by aggregating hard chain segments. At elevated processing temperatures, these crosslinks are melted or softened sufficiently to allow plastic shaping, and reforms upon cooling to give the elastic property. Hence, thermally reversible formation of physical crosslinks of the hard segments are the basis for the TPE behavior.

RESULTS AND DISCUSSION

PVC PROPERTIES

Since the 1950's, PVC has achieved and maintained a prominent position in the medical packaging and devices industry due mainly to its high performance/cost ratio. Some of the advantages commonly cited include low cost, infinite latitude in formulation that covered modulus ranging from greater than 1 GPa to less than 10 MPa, excellent optical clarity throughout this composition range, with measured haze values less than 5%. In addition, in the flexible formulations, good to excellent elastomer properties in many instances rivaling that of the natural rubber are obtained. These properties make flexible PVC one of the earliest known (but not generally recognized) thermoplastic elastomer (TPE). PVC achieves many of its remarkable properties by the possession of a small by distinct amount of crystallinity. The crystalline domains are substantially melted during processing, which allows for thermoplastic deformation and flow, but reforms upon cooling.

It is well known in the PVC industry that the plasticizer selected does not fully dissolve the crystals, upon which all properties are lost. The crystalline domains act as tie points between elastomeric plasticizer swollen amorphous main chains to confer the elastomeric properties at room temperature. Since these crystalline domains have melting temperatures upwards of 230°C, they are critical in maintaining the shape, preventing bulk flow at elevated temperatures like autoclaving. On the subject of autoclavability, one is reminded for PVC, typically only flexible, or high plasticizer formulations are autoclavable, for more rigid formulations, due to the relatively low glass transition of lower than 90°C, severe distortions due to molded in stresses are seen. Due to this "anomaly", the heat distortion property for PVC has a complex relationship with plasticizer content. Today, we see PVC being used in applications ranging from food packaging films to primary medical solution containers, to delivery tubing and catheters, to rigid housings for devices, truly an ubiquitous presence in the food and medical industry.

METALLOCENE POLYETHYLENES

Recently, metallocene catalyzed polyolefins have become commercially available. In strong contrast with the traditional Ziegler-Natta catalyst, the metallocene catalyst is the result of cyclic organo ligands coordinating on a group IIb transition element usually zirconium. The active catalytic center can be isolated and studied in detail, and more importantly, for the first time in history, the catalyst and the coordination ligand structure were found to control the polymerization activity and selectivity. This fundamental discovery is the basis of the tremendous success and the huge commercial scramble we are witnessing. From our own studies and data available from the literature, we can say the following:
- The catalyst produces extremely high yield.
- Broad monomer class can be polymerized.
- Homogeneous activity toward comonomers.
- Exquisite structural controls are possible.
- Novel polymers are emerging.

With the medical industry's concern for material cleanliness and patient safety, the extremely high yield is a welcoming attribute indeed. For polyethylene, yields as high as 40 million grams of polymer per gram of zirconium have been reported. This translates to about 25 parts per billion (ppb) of transition metal contamination in the final product if the entire amount of catalyst were left in the product. If, as expected, these catalyst yields are maintained in full-scale productions, metallocene based polyolefins will represent some of the cleanest polymers available to the medical industry.

When co-monomers are introduced to adjust the density or crystallinity of polyethylene, unlike with Ziegler-Natta catalysts, a single reaction site with uniform reactivity toward both co-monomers is present during the entire polymerization process. As a result, a very homogeneous placement of short branches across the entire molecular weight distribution is evident. Instead of multiple melting points (or, broad melting ranges) representing various crystallinity fractions, a single melting point proportional to the co-monomer content is created. The significantly lower melting point leads to easier fabrications in heat seal initiation temperatures.

One of the major benefits to the medical industry from the homogeneous structure is the drastic reduction in light scattering from larger crystallites (the higher melting fraction). Superbly clear films can be fabricated without resorting to special processing techniques or additives.

Unfortunately, the homogeneous incorporation of monomers, which led to superb clarity and flexibility, also reduces the melting point in a monotonic fashion, with reduced high temperature properties.

HIGH TEMPERATURE TPE AND ALLOYS

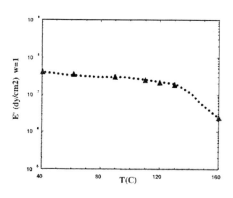

Figure 1. SEBS/PP alloy modulus vs T.

It is well known, that through polymer alloys, properties not present in the parent polymers can be obtained. For example, high melting point polypropylenes can be blended with flexible, high content ethylene vinyl acetate copolymers (EVA), to get moderately flexible, marginally autoclavable compounds. However, since the constituent components are not thermo-dynamically miscible, the overall performance was not deemed acceptable for broad usage. In a similar way, propylene and styrene based block copolymers, especially the hydrogenated mid-block polymers (SEBS), or styrene-(ethylene butene)-styrene has been used in certain autoclavable applications (Figure 1). Although the stress whitening tendencies were much reduced, the high cost of the SEBS block copolymer again hampered wide acceptance.

When polybutelene terephthalate (PBT) hard segments are co-polymerized with soft segments of polytetramethylene ether (PTMO), a multiblock thermoplastic elastomer is produced. By adjusting the ratio of PTMO in the hard segment, melting points ranging from about 150 to about 220°C can be obtained. These high performance TPE's are very tough, oil resistant, and due to their high melting points, autoclavable. Unfortunately, they tend to be opaque; thus rendering them somewhat unsuited for packaging applications.

Another class of thermoplastic polyester elastomers are based on the same PTMO soft segments, but with a hydrogenated aromatic diol (cyclohexyl dimethanol) as the hard PBT segment modifier resulted in a very minute crystallinity with a melting point of about 190°C. The material is flexible, very tough, optically transparent, and passed medical extractable testing. However, the apparent high cost has limited its use to niche autoclaving applications.

PROPYLENE BASED ELASTOMERS:

In the rubber and elastomer industry, propylene based products are well known. Both ethylene propylene rubber (EPR) and terpolymers with a crosslinkable diene monomer like 1,4-hexadiene or ethylene norbornene (EPDM) possess good high temperature elastomeric properties when crosslinked. However, the need for crosslinking and the lack of optical clarity has limited the application for medical uses.

However, propylene based elastomers with a high crystalline melting point of about 160°C could be very attractive as candidates of thermoplastic elastomers. This is the case if blocky sequences of isotactic and atactic conformations can be produced on the same polymer chain. Indeed, historically, several examples have been offered.

Early in the pioneering studies by Ziegler and Natta on coordination catalysis of olefins, it was discovered certain classes of titanocene compounds exhibited only moderate stereo regulation on the placement of methyl side group during the polymerization of propylene. The resultant polymer exhibited, not surprisingly, the elastomeric properties. However, since at the time, high isotactic yield was the aim, combined with relatively low polymer yield, this significant discovery was soon forgotten.

Later, in more systematic studies of stereo regularity control, J. Chien and co-workers at the University of Massachusetts discovered a family of higher productivity elastomeric PP catalysts. Still, lack of commercial interest hampered the full-scale development. Similarly, work at the E. I. Dupont company led to the discovery of moderate productivity catalyst, elastomeric products (Figure 2) with interesting properties. Furthermore, these Dupont researchers conducted extensive studies on the melt blending of the elastomeric PP and commercial high isotactic content PP. Since in the molten state, both types of polymer should be miscible with each other, a single phase is expected. Upon cooling, some of the crystallizable segments from the elastomeric PP could co-crystallize with the isotactic PP, thus forming the physical crosslink sites required for TPE behavior. In this way, very different mechanical properties can be achieved simply by varying the ratio of the elastomeric and isotactic PP fractions. Although research along this catalyst family continued, to-date, no commercial production of this very novel invention has been available.

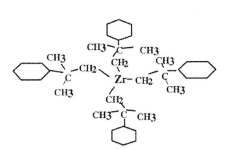

Figure 2. Dupont elastomeric PP catalyst.

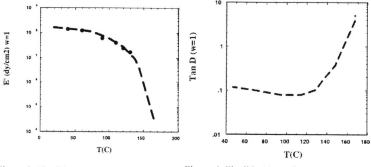

Figure 3. Flexible PP modulus.   Figure 4. Flexible PP tan δ.

In the early 1990's, based on their experience in producing amorphous polypropylene for adhesive and asphalt modification applications, the Rexene Company began semi-commercial and commercial production of the so-called flexible polyolefin (FPO[R]). And later, Rexflex[R]. This line of polymers were characterized as having previously unheard-of low modulus, as low as 7-10,000 psi (49 to 70 MPa), relatively high strength, and peak melting points of about 155°C (Figures 3, 4). This line of specialty polymers answered many of the requirements in the medical plastics where high melting point and low modulus were needed. However, the polymer was still not ideal, due to the relatively high low molecular weight and amorphous fractions in the polymer. This resulted in a very sticky and hard to handle pellets. Although difficulties in polymer manufacturing and fabrication were overcome, products made from these polymers were found to be prone to exudation and sticky, especially over time and elevated temperature storage.

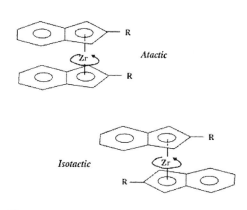

Figure 5. Rotating catalyst for stereoblock PP.

More recently, the exciting work of R. Waymouth of Stanford University showed great promise to be the most versatile catalyst system studied to-date. Basically, the catalytic behavior is controlled by the conformation of the stereo conformation of the catalyst center (Figure 5). An asymmetric indenyl metallocene ligand is coordinated on the Zirconium or Hafnium metal center. Since the two ligands are un-bridged, that is, they are free to rotate in the same plane of the their respective aromatic rings, their relative orientation controls the stereo regulation of the inserting propylene monomer. In one orientation, predominantly isotactic polypropylene is produced, the other, atactic. Hence in a very elegant way, the requirements for the alternating blocky constructions of crystalline and elastomeric segments are met.

SUMMARY

Various ways to achieve optically clear thermoplastic elastomeric behavior and high temperature properties were examined. The flexible plasticized PVC was used as a standard of comparison. To attain the attributes such as optical clarity, easy processibility, high temperature stability as well as cost for medical devices, the metallocene catalyzed thermoplastic elastomeric polypropylenes hold a great promise and bright future.

REFERENCES

1   G. H. Llinas, S. H. Dong, D. T. Mallin, M. D. Rausch, Y. G. Lin, H. H. Winter, and J. C. W. Chien, "Crystalline- Amorphous, Block Polypropylene and Non-symmetric Metallocene-Catalyzed Polymerization" *Macromolecules*, **25**, 1242-125 3, (1992).
2   C. K. Shih, A. C. Su, Poly-(α-olefin Based Thermoplastic Elastomers, 91-116, **Thermoplastic Elastomers**, N. R. Legge, G. Holden, H. E. Schroede eds. *Hanser*, New York, (1987).
3   M. Gahleitner, H. Ledwinka, N. Hafner, W. Neissl, Supersoft PP-New Properties from a Novel Molecular Structure. 283-290, Proceeding of SPO'96, Houston, (1996).
4   B. J. Pellon, Novel, High Melting Point Flexible Polyolefin Products, 401-420, Proceedings of SPO'95, Houston, (1995).

An Outlook for Metallocene and Single Site Catalyst Technology into the 21st Century

Kurt W. Swogger

*Global R&D Director, Polyethylene & INSITE Technology R&D, Dow Plastics, B-1607
Building, Freeport, Texas*

INTRODUCTION

Single site catalysts have come to be very commercially significant with the latest two catalyst families-metallocenes and constrained geometry catalysts. In 1991, Exxon announced its metallocene chemistry called Exxpol, followed by Dow in 1992 with the constrained geometry family. Both of these catalysts (see Figure 1) have a single site polymerization center and are able to incorporate a variety of alpha-olefin comonomers at high incorporation level.[1-4]

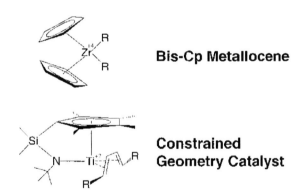

Bis-Cp Metallocene

Constrained Geometry Catalyst

Figure 1. Single-site catalysts.

Much has been written about the impact of these catalysts on the polyethylene and polypropylene industries. There are many of the major players such as Dow, Exxon, Fina, Hoechst, Mobil, Mitsui, Idemitsu, and Millennium involved in the technology race to establish competitive advantage in these fields. Less has been written about these materials in other markets such as elastomers, adhesives, and engineering thermoplastic; even though in these markets, the impact of this new chemistry may be more significant than in the more well known markets. This discussion will offer an outlook for the materials made by these catalysts in several of these markets.

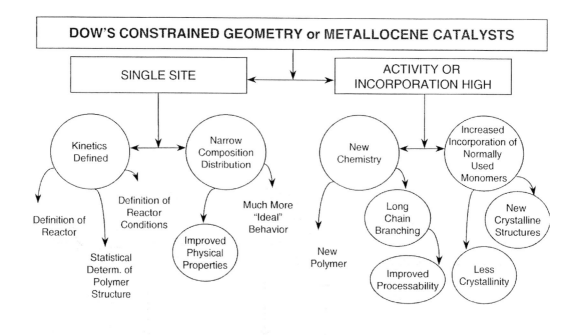

Figure 2. Use of single site catalyst for molecular architecture.

DISCUSSION

Since 1991, over two billion pounds of capacity has been installed by several companies in gas phase, slurry phase, solution phase and high pressure processes using supported and unsupported single site catalysts. Products being developed are plastomers, polyethylene, polypropylenes, elastomers, ethylene-styrene polymers, cyclic olefins and others. Alliances and partnerships have been formed or will be formed, such as Dupont-Dow Elastomers, Univation and the Dow-BP joint development, to name a few. More of these partnerships will come as the industry tries to maximize the return from these inventions by leveraging other companies and their strengths.

Why does our industry have this tremendous investment and activity? These new catalysts have high activity and incorporation rates, plus being single sited.

Figure 2 shows a schematic with the implications of these two facets of these catalysts. Polymers from these catalysts have narrow composition distribution which lend themselves to the polymer behaving more like ideal polymers. These polymers follow the polymer science rules; they act more like ideal polymers and can be modeled to predict their performance. Since they are from a single catalytic site and not an ever-changing complex mixture of different activation energy sites, they are easy to model (if you consider 30-50 kinetic expression to be solved simultaneously easy!) and use as the basis of reactor design and operation. Once a reactor is controlled, specific molecules can be designed and used as models to solve for molecular structure-physical property relationships. Knowing the kinetics allows a statistical determination of the molecular structure which is also necessary for accurate models of structure/property relationship.

Some of these catalysts (such as the constrained geometry catalyst) allow new chemistry such as long chain branching, higher levels of octene incorporation,[5] ethylene styrene copolymers plus many others. This new polymer chemistry has given a whole new array of properties which are useful and valuable for design.

With this technology, the industrial scientists can now use a molecular architecture approach to build comprehensive models and design new products. The molecular architecture concept, that of designing a polymer to meet certain performance requirements with a greatly enhanced ability, has more control of olefin chemistry from chemists to engineers and materials scientists. The whole process can be and has been modeled. A polymer can be designed to

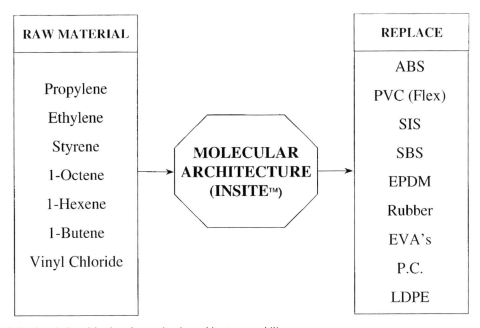

Figure 3. Product design vision based on molecular architecture capability.

meet customer performance requirements based on its materials science and then using process models-the plant conditions can be set to make that polymer.

Implications of this new chemistry include the new control of molecular structure and this modelability help extend olefin chemistry into new markets, help bring new designs of olefin molecules into existing markets quickly and greatly enhance the product development cycle.

As can be seen in Figure 3, a vision for this new capability can be shown simply by this model. The lowest cost monomers such as ethylene, propylene, styrene etc. are used as building blocks using low cost polyolefin processes to build polymers that can replace the performance of higher cost polymers such as ABS, SEBS, flexible PVC, high pressure polyethylenes and so on. This does not mean all of the applications of these higher value polymers-just some of them. That "some" will mean much more value for the practitioners of the new chemistry and some capacity issues for the incumbents.

This vision does not ignore improved polyethylenes and polypropylenes-these will be developed but will have less relative value than other markets.

A traditional development process for polyolefin took many months to years to complete. See Figure 4 for an illustration. As mentioned previously, the single site catalyst technology allow the industrial scientists to design new products using molecular architecture

Development Process

Traditional Process (6-12 Months)

Catalyst ⟶ Polymer ⟶ Products ⟶ Final Application

→ Commercialize | Yes

Time | No

New Process (1-3 Months)

(Molecular Architecture Approach)

Define Performance Requirements ⟶ Design Product ⟶ Design Process ⟶ Make Polymer ⟶ Does It Match Performance Requirement?

→ Commercialize | Yes

| No

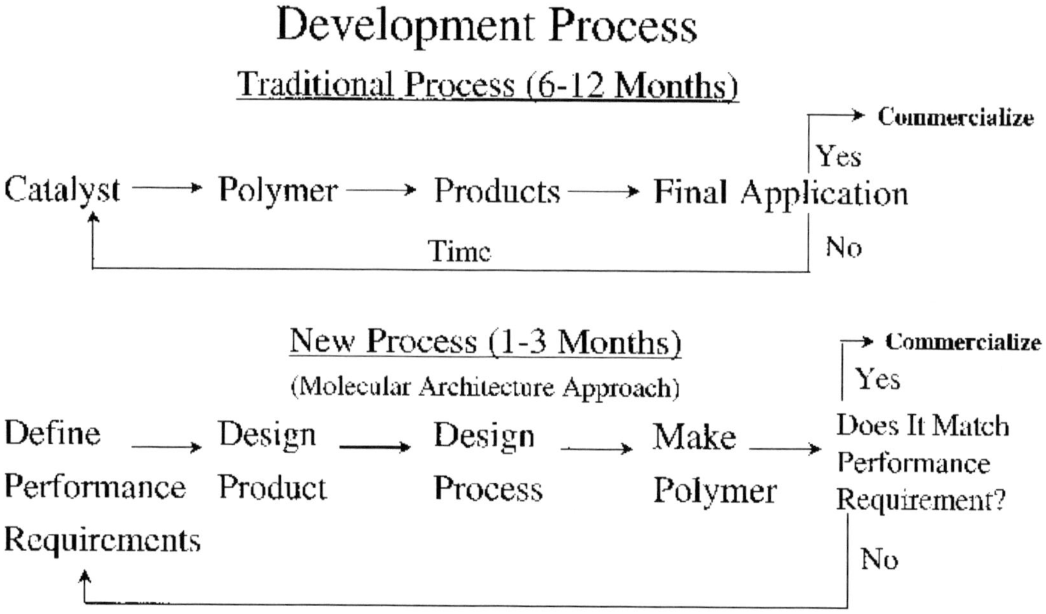

Figure 4. Development process.

approach. This greatly enhances the development process and shortens the time to develop and deliver a designed product to the customers. Figure 4 illustrates the traditional development process as well as the philosophy for the new development process using molecular architecture approaches. A polymer was developed based on a catalyst in a process; then applications were developed based on the polymer properties. Refining the polymer to meet the needs of the applications took much iteration and time. With the advent of the new chemistry allowing molecular architecture, a new approach is possible. Performance requirements are used to start the process-these being defining material and processing properties required to make the product be successful in the application. These properties are used to define a polymer structure using structure property relationships as mentioned earlier. The process conditions are established to make the polymer using reactor models. All of these models are on a computer which means the whole process takes only minutes. These models can quickly give the desired polymer or tell if it is not possible.[6-7]

This approach implies a tremendous market and customer focus. Technology is matched to a market need with specific requirements needing total definition. The high cost of technology questions by marketing has been dramatically cut-putting the focus on marketing to pick applications and customers, put together a value proposition, seeking from technology a verification or not of the feasibility of the hypothesis and doing all of this inexpensively.

Remember, you can only model what you know. Establishing the structure-property models is a major task. Once a relationship is established, it is part of the model; but, new applications often require an understanding of structures on a property not yet modeled. Over time, most desired properties will be modeled.

To summarize, my vision is that these low cost monomers can be polymerized in a low cost process to replace higher value polymers. The product development cycle will be greatly enhanced using computer models and an accelerated new product development process.

The benefits to our industry are several, including lower system costs for our customers, enhanced recyclability, lower costs and lower density. Lower system costs can come from lower density, lower material costs, improved physical properties and better conversion costs. By using some of the new materials, all olefin structures such as automobile applications, can be recycled much more readily. The lower density and downgaugeability of these materials generally allow less use of polymer per application, cutting costs as well as reducing material to the landfill.

Some examples of these new materials and their benefits are illustrative of the potential impact. Figure 5 illustrates the range of crystal morphology for ethylene/octene copolymers made possible by constrained geometry catalyst.[8] It clearly shows fringed micelle crystals are formed in high octene polyolefin elastomers and lamellar crystal for the main structure of lower octene polyethylenes. The properties of these polymers with different crystal morphology are drastically different. Even they are all made of ethylene and octene. Figure 6 shows

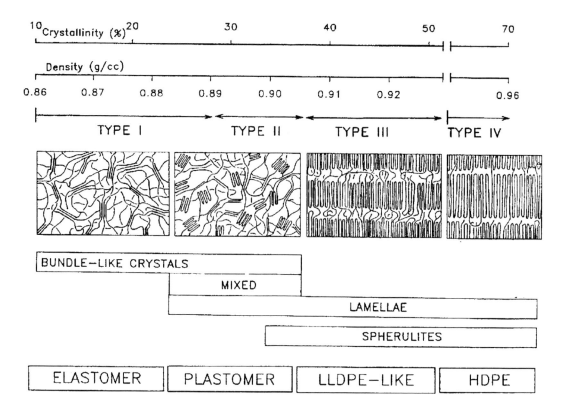

Figure 5. Range of crystal morphology for ethylne/octene copolymers made by INSITE[TM] technology catalyst.

the range of physical properties from just two monomers that is possible using these new cata-lysts. Elastomers with high levels of octene can be used as impact modifiers. New improved performance polyethylenes can be developed using copolymers with lower levels of octene.

Impact modification of polypropylene by these materials gives an improved range of properties as shown in Figure 7. With this improved range of properties, TPOs can be used even more broadly in applications such as automotive.[9]

By using the molecular architecture approach, new polyethylenes can be designed that will allow more product down-gauging. Figure 8 shows some comparisons of physical prop-erties comparing the new polymers to more established polymers.[10] Customers will take advantage of the properties to down-gauge.

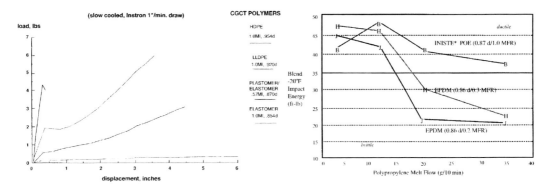

Figure 6. Load/displacement curve for ethylene/octene, HDPE, LLDPE plastomers and elastomers produced by constrained geometry catalysts.

Figure 7. Properties of TPO made by impact modification of PP with INSITE[TM] technology POE vs. conventional EPDMs. (blend: 70% PP + 30% elastomer).

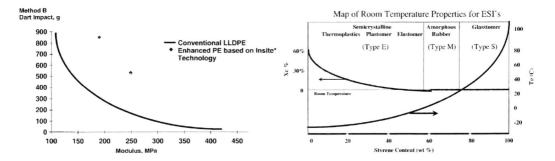

Figure 8. New performance combination (high dart impact and high modulus) for enhanced PE produced by INSITE[TM] technology.

Figure 9. Glass transition temperature and crystallinity of type E, type M and type S ethylene/styrene interpolymers (ESI).

New chemistry which is now possible includes ethylene/styrene interpolymers. Figure 9 show three types of materials made from these monomers and some of their distinguishing characteristics.[11-12] Figure 10 shows a list of new enhanced properties. These new polymers will overcome performance deficiencies in some applications and allow cost reduction in others. Using the constrained geometry catalyst to combine high ratios of styrene to ethylene creates unique polymers and shows the power and utility of these new catalysts.

By the year 2000, there will be at least 7 billion pounds of capacity for metallocenes and constrained geometry catalyst-based products. There will be additional capacity for EPDMs,

- Unique Material - 0 to Approx. 50 Mole % Styrene
- Surface Characteristics- Adhesion, Abrasion
- Toughness
- Weatherability
- Barrier
- Rheology
- Compatibility
- Chemistry- Sulfonate, Hydrogenate…
- Dampening Characteristics
- Stress Relaxation and Recovery

Figure 10. Top ten ESI attributes.

E/O elastomers, plastomers, new polyethylene and polypropylenes, cycle olefins, and ethylene/styrene interpolymers.

There will be single site capacity in gas, solution, slurry and high pressure units around the world. These materials will be high performance, easy to model, and have lower cost. They will be making inroads into major non-olefin markets such as automobiles, rubbers, and elastomers.

While there are currently two leaders in the industry, many others will be able to participate in various segments of the market. The race will be very quick to establish franchises and will continue to pressure each party to keep on track. Leadership, money, survival are all parts of this desire to push these programs.

What will slow the race? Intellectual property disputes will be targeted for quite a while. Which companies are serious and not serious about these positions? Can we get enough highly skilled people to do the necessary development? These are critical issues that, depending on how a company responds, will determine success or failure.

The metallocene and constrained geometry catalyst systems produce polymers that have valuable properties. There is no longer any doubt. They will have a profound impact on the industry not only in materials, but who will be the technology leaders and even what the structure of the industry will be. This technology change truly will transform the plastics world.

REFERENCES

1   K. W. Swogger, "The Material Properties of Polymers Made from Constrained Geometry Catalyst", Proceedings of the Second Int'l Business Forum on Special Polyolefin, SPO '92, pp. 115 (1992).

2   B. A. Story and G. W. Knight, "The New Family of Polyolefins from INSITE Technology", Proceedings of Metcon '93, pp. 112 (1993).

3   C. S. Speed, B. C. Trudell, A.K. Mehta, F.C. Stehling, "Structure/Property Relationships in Exxpol[TM] Polymer", Proceeding of Polyolefins VII International Conference, pp. 45 (1991).

4   S. P. Chum, C. I. Kao and G. W. Knight, "Structure and Properties of Polyolefin Plastomers and Elastomers Produced from the Single Site, Constrained Geometry Catalyst", Proceeding of Polyolefins IX International Conference, pp. 471 (1995).

5   G. W. Knight and S. Lai, "Dow Constrained Geometry Catalyst Technology: New Rules for Ethylene/Alpha-Olefin Interpolymers-Unique structure and Property Relationships", Proceedings of Regional Polyolefin Technology Conference, SPE 10 RETEC '93 (1993).

6   K. W. Swogger, "Expanding Polymer Structure Control and Design Capabilities Through INSITE* Technology", Proceedings of SPO '94 (1994).

7   S. P. Chum and J. Ruiz, "Using Polymer Structure/Property Model for Molecular Architecture Product Design", Proceedings of 1997 Aspenworld Technology Conference (1997).

8   S. Bensason, J. Minick, A. Moet, S. Chum, A. Hiltner and E. Baer, "Classification of Homogeneous Ethylene-Octene Copolymers Based on Comonomer Content", *J. Polym. Sci., Part B: Polym. Phys.*, **34**, 1301-1315 (1996).

9   H. C. Silvis, R. C. Cieslinski, D. J. Murray and S. P. Chum, "The Use of New Polyolefin Elastomers for Impact Modification of Polypropylene", SAE Technical Paper Series #950559 (1995).

10   K. W. Swogger and S. P. Chum, "Enhanced Polyethylene Using INSITE* Process Technology", Proceedings of Global Metallocene Technology Conference Metcon '96 (1996).

11   K. W. Swogger and S. P. Chum, "Material Properties of Ethylene Styrene Interpolymers", Proceeding of Metallocene Technology Conference Metcon '97 (1997).

12   Y. W. Cheung, M. J. Guest, "Structure, Thermal Transitions and Mechanical Properties of Ethylene/Styrene Copolymers", Proceeding of Antec '96 (1996).

World Olefins Markets: Issues Influencing Availability and Pricing

Robin G. Harvan
Bonner & Moore Associates, Inc.

BACKGROUND

Plastics is a generic term used to describe large hydrocarbon-based molecules that, when heated, soften enough that they can be shaped. The word polymer is often used interchangeably with plastics, though its technical definition encompasses other categories of products. What is important is not the definition of polymer or plastic, but rather the knowledge that polymers consist of smaller molecular building blocks called monomers. Put simply, therefore, a polymer is made up of monomers.

Monomers are composed of carbon and hydrogen, and fall into two basic groupings which relate to the manner in which the carbon and hydrogen atoms are arranged. Aromatic monomers are circular in arrangement, while olefin monomers are linear. We mention all this simply because we will be addressing only olefin monomers in this discussion.

There are three basic olefins: ethylene, from which polyethylene is made; propylene, from which polypropylene is made; and butadiene, from which an array of plastics materials such as acrylonitrile-butadiene-styrene (ABS) resin and synthetic rubbers are made.

Olefins are made by breaking long-chained hydrocarbon molecules into smaller lengths by the application of heat and pressure in the presence of a catalyst. There are several processes for doing this, all of which carry the generic name cracking. It is from this generic term that we derive the name steam cracker for the production unit in which olefins are manufactured.

While the materials being cracked, the so-called "cracker feedstocks," are several with variations that are both regional and situation in aspect the products and byproducts of the olefins process are more or less identical around the world. This similarity also tends to result in manufacturing costs and prices being relatively comparable, once region-/site- specific

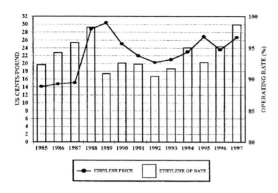

Figure 1. Relationship between ethylene operating rate and price. United States Gulf Coast.

factors such as transportation and scale-economies are standardized. All this only further supports the statement that the olefins business is a truly worldwide one.

ANALYSIS AND FORECAST

As may already be apparent, we contend that a clear link exists between changes in olefins production operating rates, and prices for those olefins. This linkage is such that olefins prices will tend to follow movement in utilization. The validation of this premise is shown in Figure 1, which presents data on average annual pricing and average annual operating rates for the U.S. ethylene market for the period 1985 to 1995. The nature of the olefins business is such those ethylene market conditions tend to provide directionality to all olefins activity. The reason for this is quite simple: most propylene and butadiene are derived as byproducts of ethylene manufacture. With this being the case, the movements in propylene and butadiene operating rates and prices will tend to mirror those of ethylene.

From this graph, one can see that prices tend to rise once ethylene utilization rates move past 94 percent. Just such a situation occurred in the 1985 to 1988 timeframe. Prices remained relatively stable advancing, though, at less than the rate of inflation through early 1987. At that time, operating rates moved upwards to average 95.2 percent for the year. Concurrently, ethylene prices moved up abruptly, advancing from an average price of 14.8 cents/pound in 1986 to 15.1 cents/pound to 1987.

It should be noted that these average prices imply a high/low configuration for the year. And 1987 was no exception. Ethylene prices rose steadily all year, finishing in December at 17 cents/pound. The following year saw utilization push past 98 percent, and prices almost doubled to a 28.8 cents/pound average.

In 1989, operating rates dropped sharply to an average of 91 percent, as capacity additions and a slowdown in ethylene derivative demand-itself a sign of the approaching recession-caused a market reversal. However, while utilization rates moved down, the ethylene price coasted higher to an average of 30.4 cents/pound. The word "coasted" is particularly appropriate, because we believe that it effectively describes the dynamics in play; namely, that once operating rates do rise, there appears also to be a slight lag in the price fallback.

We attribute this price momentum to the fact that ethylene is very much a just-in-time-delivery type of product. Ethylene derivative manufacture is a continuous process. As such, producers want assurance that not only is capacity available to make ethylene, but also those on-hand stocks of ethylene are sufficient. It is typical in the olefins industry-and for ethylene in particular-to view sufficient on-hand stocks as a number equal to one week of industry production (this unit of measurement being average quarterly capacity divided by the number of days in that quarter).

As such, both buyers and sellers pay especially close attention to inventory levels. This is particularly so in time of high utilization, as on-hand inventories tend to be depleted at those times. This often leaves the market with a sense of perceived shortage despite there being sufficient capacity to meet real demand.

Still, operating rates below 94 percent will lead to lower prices. The effect will tend to be exacerbated by the prolonging of the downturn. This can be seen in the 1989 to 1993 period, when the U.S. economy was deeply in a recession. Here we find prices slumped from a 30.4 cents/pound average in 1989 to an average of just 20.3 cents/pound in 1992. This 30-plus percentage drop in prices was a primary causal factor in the slowdown and virtual cessation of steam cracker construction during this period of time.

And, as we would expect, prices did not begin to rise until the same happened to operating rates. We see that from 1993 on, as operating rates moved towards and then past 94 percent (as they did in 1994), ethylene prices were also on the rise.

Why is it, then, that prices stayed on the rise in 1995, a year when the average operating rate retreated? The answer to this is the one stated above: prices can rise if the perceptions or the reality of on hand inventory levels implies a shortage. In August 1994, Exxon Chemical and Shell Chemical both had steam cracker explosions which put these units out of action for six months. During this period, inventories were drawn well down. In the resultant environment, most buyers and sellers believed another accident, were it to occur, would leave the industry unable to meet demand. This being the case, ethylene was perceived to be at a premium; this situation was reflected by the rise in the 1995 average price for ethylene to some 27 cents/pound.

In 1996, another anomaly occurred, as utilization rebounded to 95.2 percent, while prices slipped to an average of 23.6 cents/pound. Here, we find the reverse of 1995. Capacity additions-both new and repaired brought so much output into operation that avails rose sharply ... sharply enough, in fact, that the market shifted briefly in favor of buyers. This situation led to a fallback in prices.

Looking to 1997, prices and operating rates are both on the rise, as more normal conditions become extant. Ethylene should average 26.7 cents/pound in a market where steam cracker utilization is 98.7% percent.

What, then, does the future hold for olefins prices?

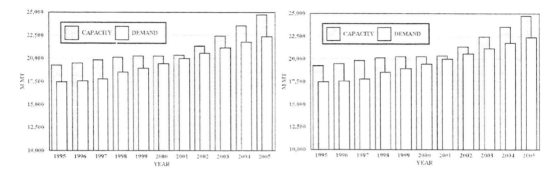

Figure 2. North American ethylene supply/demand.   Figure 3. West European ethylene supply/demand.

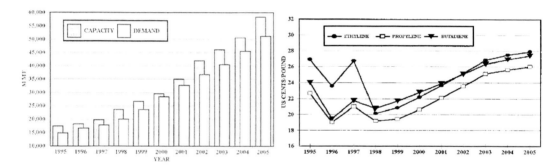

Figure 4. Asia-Pacific ethylene supply/demand.   Figure 5. Olefins price history and forecast. US Gulf Coast.

To answer this question, we have developed supply/demand forecasts for ethylene for the three major producing and consuming regions of the world: North America, West Europe, and Asia-Pacific. We do this not to minimize the importance of South America, Central and Eastern Europe, or the Middle East/Africa. In point of fact, South America holds more promise for growth than West Europe. However, for the period under study, 1995 to 2005, the three regions we address are the most important for olefins.

What we see is that ethylene operating rates will likely move lower in the next twelve to eighteen months. First, let us clarify that ethylene demand is itself subject to cyclicality. We mentioned how demand weakened in the 1990 to 1993 recession. It is our further belief that the U.S. economy will go into another recession before the end of the decade, thereby reducing demand at a time when a significant volume of new capacity will be coming into operation.

Figure 2 shows that in North America, ethylene capacity will rise by some 3.5 MM metric tons between 1997 and 1999. At the same time, demand should rise far less; this conclusion is certain whether or not a recession occurs. With supply growing more rapidly than demand, operating rates should be moving lower during this period. And, as seen from the past, successive years of retreating operating rates have triggered price fallbacks.

Figure 3 and Figure 4 show similar scenarios likely for West Europe and Asia-Pacific, respectively, though the degree of the capacity overhang is less. Nevertheless, the effect should be sufficient as to cause olefin prices also to retreat in these regions.

How do we see prices moving during the forecast period? Figure 5 presents our forecast of current dollar prices for ethylene, propylene, and butadiene during the 1995 to 2005 timeframe.

We expect prices to fall in 1998 and to stay somewhat depressed through late 1999. Thereafter, prices should begin to recover, as we project that incremental demand growth will be gradually absorbing existing capacity.

CONCLUSION

Olefins prices tend to move directionally with industry operating rates. We have seen that between 1985 and 1997, two pricing peaks occurred at a time when industry utilization moved past 94 percent. With this being the case for operating rates at or above 94 percent, it follows that operating rates moving below 94 percent will tend to pull prices down, also. As operating rates are projected to retreat in the 1997 to 1999 timeframe, we think olefins prices will do the same.

World Polypropylene Demand

Russ DeLuca
Amoco Polymers Inc. Alpharetta, GA 30005-3914, U.S.A.

NEW PRODUCT GRADES

Polypropylene can take on many "forms" when we consider its performance characteristics. It can be rigid and durable when needed for houseware or automotive interior applications. It can offer specific barrier properties and flexibility for packaging. It can also provide comfortable but wear resistant performance for carpeting.

Further, the performance of polypropylene in 1997 is significantly improved over products of 5-10 years ago. Continual improvements in process and catalyst technology as well as additives have yielded these improvements. The outlook is for this trend to continue. We are seeing tougher and stiffer reactor products as well as a myriad of improved additives which enhance performance. These new additives can range from clarifiers to improve functional transparency to new elastomers to increase toughness. Beyond all this is the prospect of metallocene catalyst providing a new range of performance. To date, there has been limited commercial impact but this may change over the next five years.

It is easy to see how the combination of an essentially versatile performing polymer, with broad based technology improvements, yields a powerful foundation for growth.

POLYPROPYLENE VALUE EQUATION

This driver is easy to understand. Polypropylene is the lowest density commodity polymer thus giving it a major head start in the economic equation for materials evaluation.

Polypropylene also is quite versatile in fabrication options. It is readily extruded in many ways; fiber, film and sheet. It can be thermoformed at high speed and readily blow-molded. It is injection molded in a wide array of applications. Developments in non-wovens and thin wall-molding extend the options.

Thus, the material engineer has a cost effective product that can be fabricated in many ways. While we would anticipate continual improvement in fabrication, it is not likely to provide major new incentive for growth by itself.

POLYPROPYLENE PROCESS TECHNOLOGY

Technology developments have moved ahead rapidly in both polymer process and catalyst areas. This has driven the capital and operating cost per ton down over time thus providing a basis for continuing investment in new capacity. For example a world scale plant today is 250,000 tons per year where it was 125,000 tons in 1990 in the developed regions of the world.

Given the status of technology today we would project slower improvement in cost over the next ten years compared to the last ten. First, as the scale of plants increases the next increment provides less benefit. Next the major catalyst (metallocene) focus is targeted on property improvements not cost reduction. These factors, however, will not inhibit the steady increase in capacity to meet demand.

Propylene availability could be a factor in some regions. Propylene derivatives are growing faster than ethylene derivatives; thus, creating a possible barrier where the prime source of propylene are crackers. This is not likely to impact North America, but could have impact in parts of Asia and Europe.

Thus, we can see that the three key drivers of growth are in place for the future. These now provide the basis for the marketplace to continue the trend of continually growing our existing applications and developing new ones.

Polypropylene is used in a wide array of applications. Some of the key areas where we see increased penetration include;
- automotive exterior fascia
- food packaging
- bottles
- non-wovens
- prepared food containers

Certainly we will see many new applications appear over the future years; the combination of performance versatility and value assure this.

There is a further very important aspect to overall polypropylene growth. That is the development of many emerging regions of the world. In each of these areas, the plastics consumption per capita is a fraction of that in U.S. or Western Europe. This usage will grow steadily as the economies in these regions mature. The key regions for this growth will be the Mercosur trade region of South America, eastern Europe, eastern Mediterranean and large parts of Asia including China, India and the array of countries in southeast Asia.

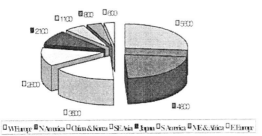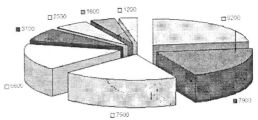

Figure 1. Polypropylene global demand: 1996 in thousand metric tons.

Figure 2. Polypropylene global demand: 2005 in thousand metric tons.

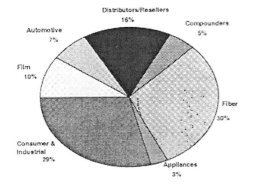

Figure 3. US polypropylene demand; 1996. consumption by segment.

This is the stage that is set for the future growth of polypropylene.

We will now address the growth projection by region The breakdown of our growth projections by region are shown in Figures 1 and 2 for 1996 and 2005.

North America

North American demand was approximately 5 million tons in 1996 and is projected to be 8.0 million tons in 2005; a 6 per cent average annual growth rate. The current end uses for polypropylene are shown in Figure 3.

We expect growth in automotive and rigid packaging to be somewhat higher than other segments, but this will not dramatically alter this data by 2005. The demand by product family are shown in Figure 4.

Western Europe

This region has the highest current and projected future demand at 5.6 and 9.2 million tons (2005).

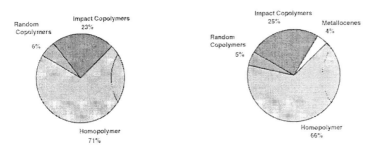

Figure 4. US polypropylene demand. Consumption by product family. Left 1996, right 2005.

The breakdown of end uses is shown in Figure 5. You will note that there are some significant

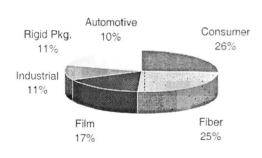

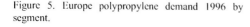

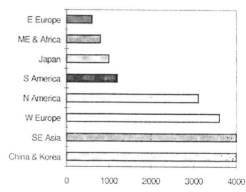

Figure 5. Europe polypropylene demand 1996 by segment.

Figure 6. Increase in polypropylene demand 1996-2005.

Table 1. Asia polypropylene demand in thousand metric tons

	1996	2005
China	2615	5515
South Korea/Taiwan	1336	2210
India	327	1637
Southeast Asia	1651	4025

differences from North America. These are due in some instances to consumer preferences (more carpet sales in N.A.) and in other cases to the emphasis for specific materials in application development.

Asia

Asia is of particular interest as 9.0 million tons of increased demand will occur between now and 2005 in this region (see Figure 6). To a great extent, this demand is driven by China. This country will be a net importer of polypropylene through the period to 2005. South Korea will be a major exporter primarily to China during this period and southeast Asia will be overall relatively balanced with some cyclical periods. However as a whole, Asia will have a balanced supply/demand picture in the next several years, due to the large number of new projects currently underway. Of course, individual countries within Asia will continue to remain net importers of polypropylene (Vietnam being an example in addition to China). The specifics of Asia demand are in Table 1.

PVC Markets, Today and in 2020

Donald Goodman

Occidental Chemical Corporation, Pottstown, PA 19464

PVC TODAY

Because of PVC's combination of versatile performance and low price, PVC will continue to grow at about 7±3% for the decade. The variability will result from new production capacity relative to world economies. The areas of highest growth are in global construction activities and export to developing nations for all applications. There have been and continue to be a number of environmental challenges that have been irritating to outright menacing of particular PVC markets. The following outline will attempt to highlight these drivers and barriers to PVC markets.

In the emerging and survival economies, there are opportunities of infrastructure development that are favorable for PVC. For delivery of clean water systems, for example, there will be high demand for PVC pipe. Similarly, for sewage treatment, transport, and general sanitation improvement, much PVC will be used. Inexpensive PVC modular housing is gaining popularity. The increased use of PVC will prevent, or significantly reduce, deforestation and subsequent soil loss. The use of PVC for basic irrigation, without overuse of water, is an established practice. The point is that for both domestic and overseas construction applications, there will continue to be high demand for PVC.

PVC has been winning many moderate performance/low cost plastics applications. This should continue for a number of years. There is expected to be an impact of metallocene catalyzed polyolefins on flexible PVC markets in the next two decades.[1] Higher performance PE and PP can be tailor made with new metallocene complexes. As these catalysts become less expensive and as the properties improve, then some impact on PVC flexible markets may be expected. This impact should be gradual but substantial (up to 15%) over the next 20 years.

On the environmental front, PVC has remained steady and equal to all the challenges. The marketing difficulty is that the perceived rather than the real problems are being believed. The PVC industry has undergone a revolution in recent years from a closed private aspect to a transparent and open industry. Measurements of dioxin and vinyl chloride, for example, in products, byproducts, and emissions are known to regulators, to antagonists, and most importantly, to industry itself which continues to strive to further improvements in reducing toxic emissions.

Even as the products and processes are becoming cleaner, the propaganda attacks to PVC have become more pointed. The Dusseldorf airport fire created problems for PVC markets because PVC was mistakenly blamed as a fuel in the fire, and injury from dioxins was supposed to have hurt people. These claims were later proved to be untrue, and PVC was proved to have had very little involvement in the fire. The adverse publicity during this investigation, however, turned out to be harmful to vinyl markets. Similarly, claims of lead chalking on vinyl slat blinds gave the industry unnecessary heartburn because lead stabilizers are not used for these applications at all in the United States.

The environmentalists have invented a "poison soup" concept around PVC. Even as the PVC resin comes out clean and environmentally friendly, claims are made against phthalate plasticizers used to make flexible PVC; tin stabilizers are a problem in Sweden although they are approved in the United States; and toys containing plasticizers and possible heavy metal colorants are all under attack. These allegations all have an impact on markets, particularly when competing materials continue the allegations.

Vinyl has had a long history of defending itself against environmental claims starting in the mid seventies with the vinyl chloride toxicity problems. Since then the industry has had considerable success in managing issues from fire science to concerns about the use of heavy metals in compounding. Despite all these issues and problems, the business impact on markets has been marginal.

PVC TOMORROW

In the Vinyl Institute[2] report "2020" prepared by Dr. Gary Gappert in 1996,[3] four futuristic scenarios are presented: (1) vinyls boom in second and third worlds; (2) there is a major replacement of vinyl in the 25 years; (3) there is selective sunsetting of vinyl products; and (4) PVC is saved by advancing technology and is thriving. These scenarios are analyzed against four economic conditions (a) continuation of existing economic trends; (b) acceleration of new trends; (c) maturation of existing trends; and (d) there are restrictions on the use and manufacture of vinyl.

The scenarios are based on adverse political pressure to decrease PVC production or participation in particular markets. The other major variable is the level of innovation or technical change and improvement. The prediction is made that the industry could grow to as much as $61 billion by 2020. Adverse political policies and technical stagnation may restrict that growth to less than $23 billion in 2020. The $38 billion potential difference between the most favorable and least favorable scenarios requires serious attention to political policy intervention and to advanced technology development.

Among the areas listed in "2020" that are favorable for technical development are:[4] global warming and rising ocean levels; deforestation and rising cost of wood; worldwide urbanization and housing demand; cultural changes and regional conflicts; the need to conserve fossil fuel supplies by designing more energy efficient vehicles; health care revolution; communications revolution; and environmental requirements.

The PVC industry was given a wake-up call by Mr. Bill Patient, CEO of Geon, at the World Vinyl Forum[5] to develop the technology so that the PVC industry will continue to grow and prove to be a robust sustainable plastic into 2020 and beyond.

REFERENCES

1   "The Impact of Metallocenes on PVC," R.B. Wilson, Jr., presented at the First World Vinyl Forum.
2   Vinyl Institute, 65 Madison Ave., Morristown, NJ 07960, Tel: 973-898-6699.
3   "2020: Progress. Challenge. Prospects for the next Quarter Century," commissioned by the Vinyl Institute, April 1996.
4   Ibid., page 37.
5   First World Vinyl Forum, Akron, OH, Sept. 9, 1997.

Index

A
ABS 293
activation energy 29, 97, 212, 271
agricultural films 137
amorphous 9, 23, 37, 112, 147
      agent 130
appearance 67
atactic 9, 23, 29, 37, 187, 279
autoclaving 275
Avrami
      equation 249
      kinetics 263
       theory 157, 265

B
bags 137
barrel 73
birefringence 178, 223
birefringent 148
blends 17, 27, 29, 38, 48, 53, 62, 77, 129, 164
blow molding 87, 117, 122
blowing agent 157
branches 5, 261
branching 193, 217, 232
brittleness 29
bubble 118, 122
      stability 204

C
can liners 137
capillary flow 223
cell structure 159
chain 13
chromium 1
clarity 47, 147

cobalt 1
coextrusion 129
compatibility 17
conductivity 165
conformation 167, 180, 279
contamination 275
cooling 247
copolymers 6, 31, 164, 194, 209, 258
      distribution 77
copolymerization 10, 239
Cox-Merz rule 234
CP/MAS 178
critical stress 227
crosslinking 157, 193
crystalline 23
      modifications 29
crystallinity 94, 107, 264
crystallites 105
crystallization 18, 54, 112, 117, 150, 177, 239, 257, 263
      kinetics 168
cushioning 160
cyclic loading 78

D
damping factor 78
debonding 38
defects 102
degradation 194
die 64, 224
diffusion 54, 159
DMS 178
Doi-Edwards model 233
driving force 13

DSC 15, 26, 30, 55, 87, 103, 113, 130, 135, 170, 178, 183, 248, 252, 264, 266
dynamic modulus 82

E
elastic recovery 172
elasticity 41
elongation 20, 132, 171
endotherm 106
EPDM 77, 279
EPR 279
equipment design 154
ethylene 1
ethylene/styrene interpolymer 37
EVA 53, 86, 94, 163, 278
exotherm 267, 272
expansion 157
extensional
      flow 224
      viscosity 217
extractables 148
extruder 48, 61, 138, 154, 177
extrusion 17, 69, 86, 97, 170, 197, 201, 223

F
fatigue 78
fiber 168, 177
film 17, 38, 69, 85, 111, 117, 122, 129, 168, 177, 201
flexural modulus 27
Flory equation 240, 258
foam 38, 157
fold surface 261
food 93, 137
free energy 261
fresh-cut produce 85
FTIR 183
      microscopy 26
fusion 189

G
gas permeability 47
gelation 150
GPC 6, 18, 236
growth rate 263

H
hafnium 9
half-titanocene 12
haul off speed 132
haze 95
head-to-head insertion 101
heat
        of fusion 88, 261
        resistance 42
        transfer 54
Henry's Law 158
heterogeneous 5
HIPS 29
Hoffman-Lauritzen equation 250
homogeneous 5
hot tack 55
housing 303
hysteresis-loop 78

I
immiscible 17
impact 28, 50, 89, 138
impurities 23
injection molding 41, 81
interaction 19
interpenetrating network 81
IR dichroism 113
iron 1
isomorphism 190
isostrain 117
isostress 117
isotactic 9, 101, 279

K
Kaminsky 1
Kraton 29

L
lamellae 23, 32, 102, 251, 257
lamellar thickness 115, 240

M
markets 283
masterbatch 62
medical 93, 137, 275
melt 23, 129, 170, 177
    fracture 223
    index 69, 138, 204
    instability 97
    strength 201
    temperature 62
melting 18, 190
    behavior 69
    instability 74
    point 31, 58, 95, 102, 258
methylaluminoxane 1
microscopy 30
Mie-theory 148
military helmets 160
miscibility 27
modeling 38
modified atmosphere 86
modulus 19
molecular
    stiffness 272
    weight 77
morphology 3, 29, 38, 41, 111, 148, 247

N
nanodispersion 148
nanoscale 25
Newtonian flow 95
nickel 1
NMR 112, 231

NRA 29
nucleating agent 27, 148
nucleation 131, 158, 244, 257, 269
nuclei concentration 247

O
olefins 293
operation variables 154
optical
    clarity 275
    properties 48, 101
orientation 32, 90, 116, 177
output 62
oxygen 86
Ozawa exponent 265

P
packaging 53, 85, 93, 137, 147, 275, 299
packing 167
PBT 278
permeability 85
pipe 303
plasticizer 277
platelets 23
polydispersity 1, 13, 18, 193, 232
polyethylene 17, 41, 47, 53, 62, 77, 86, 93, 111, 121, 129, 137, 153, 161, 183, 202, 209, 234, 239, 257, 277, 283, 293
polymerization 2
polymorphism 29, 167
polyolefins 9, 111, 157, 183, 258, 277
polypropylene 10, 18, 23, 77, 101, 147, 164, 167, 177, 183, 194, 263, 278, 283, 293, 299
polystyrene 29, 37, 147
pressure 50, 65
prices 295
printability 41
processability 153, 203
propylene 13
PTMO 278
pumping capacity 64

PVC 277, 303

Q
quality 67

R
Raman 112
refractive index 42
relaxation
          spectrum 193
          studies 17
respiration 86
rheology 38, 62, 96
rheometer 80, 131, 169, 232
rotation 10
rubber 293
rubbery phase 114

S
SAXS 112
screw 61
          design 69
seal 54
sealability 41
sealing strength 91
SEBS 278
SEC 193
segregation 23, 25
SEM 158
shear rate 213
shear thinning 217
sheets 168
shrinkage 171
silica 2
SIST 103
slit-tapes 168
slurry process 47
sol-gel transition 149
solidification 38
spherulite 23, 33, 147, 247, 252, 257
spinning 168, 177

stabilization 195
stereochemistry 9, 101
stereomodifiers 147
stereoregularity 101, 188
sterilization 275
stiffness 27, 88, 150
storage 85
strain hardening 108, 235
stress analysis 178
stress-strain 78
styrene 10
supercooling 217, 251, 258
surface 67, 98
syndiotactic 12, 29, 187

T
TDA 178
tear
          resistance 89
          strength 51, 123, 138
TEM 113
tenacity 171
tensile
          modulus 88
          strength 50, 132, 178
texture 116
T_g 37, 276
thermoformability 41
tie-molecules 261
titanium 9
          dioxide 62
torque 134
toughness 27, 47
TPE 276
transcrystallization 112, 118
transparent 148
triisobutylaluminium 2
tubing 97

U
unsaturations 186, 195

V
vinyl 186
vinylidene 186
viscoelastic studies 93
viscosity 80, 134, 139, 195, 210, 233

W
WAXD 169, 178
WAXS 31, 112
wetting 54

Y
yield stress 108
yielding 107
Young's modulus 82, 117, 133

Z
Ziegler-Natta 1, 17, 69
zirconium 9